Math for Life

Crucial Ideas You Didn't Learn in School

Math for Life
Crucial Ideas You Didn't Learn in School

Jeffrey Bennett

ROBERTS AND COMPANY PUBLISHERS
Greenwood Village, Colorado

Math for Life: Crucial Ideas You Didn't Learn in School

Roberts and Company Publishers
4950 South Yosemite Street, F2 #197
Greenwood Village, CO 80111 USA
Telephone: (303) 221-3325
Facsimile: (303) 221-3326
Internet: www.roberts-publishers.com
Electronic Mail: info@roberts-publishers.com

Ordering Information
Order online at www.roberts-publishers.com.
Within the USA, telephone us at 1-800-351-1161.
Outside of the USA, telephone us at 001 303-221-3325.
Fax your order to 303-221-3326.

Editing: Joan Marsh, Lynn Golbetz
Composition and design: Side By Side Studios

Front cover photo credits:
Solar field: ©Pedro Salaverria/Shutterstock
Charlotte map:©Tupungato/Shutterstock
Texting while driving: ©George Fairbairn/Shutterstock
National debt clock: ©Clarinda Maclow

ISBN: 978-1-936221-43-1

Library of Congress Cataloging-in-Publication Data

Bennett, Jeffrey O.
 Math for life : crucial ideas you didn't learn in school / Jeffrey Bennett.
   p. cm.
 Includes bibliographical references and index.
 ISBN 978-1-936221-43-1
 1. Mathematics--Popular works. I. Title.
 QA93.B444 2011
 510--dc23
                    2011031476

10 9 8 7 6 5 4 3 2 1

Table of Contents

Preface

The housing bubble. Lotteries. Cell phones and driving. Personal budgeting. The federal debt. Social Security. Tax reform. Energy policy. Global warming. Political redistricting. Population growth. Radiation from nuclear power plants.

What do all the above have in common? Each is a topic with important implications for all of us, but also a topic that we can fully understand only if we approach it with clear quantitative or mathematical thinking. In other words, these are all topics for which we need "math for life"—a kind of math that looks quite different from most of the math that we learn in school, but that is just as (and sometimes more) important.

Now, in case the word "math" has you worried for any reason, rest assured that this is not a math book in any traditional sense. You won't find any complex equations in this book, nor will you see anything that looks much like what you might have studied in high school or college mathematics classes. Instead, the focus of this book will be on what is sometimes called *quantitative reasoning*, which means using numbers and other mathematically based ideas to reason our way through the kinds of problems that confront us in everyday life. As the list in the first paragraph should show, these problems range from the personal to the global, and over everything in between.

So what exactly will you learn about "math for life" in this short book? Perhaps the best way for me to explain it is to list my three major goals in writing this book:

1. On a personal level, I hope this book will prove *practical* in helping you make decisions that will improve your health, your happiness, and your financial future. To this end, I'll discuss some general principles of quantitative reasoning that you may not have learned previously, while also covering specific examples that will include how to evaluate claims of health benefits that you may hear in the news (or in advertisements) and how to make financial decisions that will keep you in control of your own life.

2. On a societal level, I hope to draw attention to what I believe are oft-neglected mathematical truths that underlie many of the most important problems of our time. For example, I believe that far too few of us (and far too few politicians) understand the true magnitude of our current national budget predicament, the true challenge of meeting our future energy needs, or what it means to live in a world whose population may increase by another 3 billion people during the next few decades. I hope to show you how a little bit of quantitative reasoning can illuminate these and other issues, thereby making it more likely that we'll find ways to bridge the political differences that have up until now stood in the way of real solutions.

3. On the level of educational policy, I hope that this book will have an impact on the way we think about mathematics education. As I'll argue throughout the book, I believe that we can and must do a much better job both in teaching our children traditional mathematics—meaning the kind of mathematics that is necessary for modern, high-tech careers—and in teaching the mathematics of quantitative reasoning that we all need as citizens in today's society. I'll discuss both the problems that exist in our current educational system and the ways in which I believe we can solve them.

With those three major goals in mind, I'll give you a brief overview of how I've structured the book. The first chapter focuses on the general impact of societal attitudes toward math. In particular, I'll explain why I think the fact that so many people will without embarrassment say that they are "bad at math" was a major contributing factor to the housing bubble and the recent recession; I'll also discuss the roots of poor attitudes toward math and how we can change those attitudes in the future. The second and third chapters provide general guidance for understanding the kinds of mathematical and statistical thinking that lie at the heart of many modern issues and that are in essence the core concepts of "math for life." The remaining chapters are topic-based, covering all the issues I listed above, and more; note that, while I'd like to think you'll read the book cover to cover, I've tried to make the individual chapters self-contained enough so that you could read them in any order. Finally, in the epilogue, I'll offer my personal suggestions for changing the way we approach and teach mathematics.

As an author, I always realize that readers are what make my work possible, and I thank you for taking the time to at least have a look at this book. If I've convinced you to read it through, I hope you will find it both enjoyable and useful.

Jeffrey Bennett
Boulder, Colorado

1

(Don't Be)
"Bad at Math"

Nothing in life is to be feared. It is only to be understood.
— **Marie Curie**

Equations are just the boring part of mathematics.
— **Stephen Hawking**

Let's start with a multiple-choice question.

Question: Imagine that you're at a party, and you've just struck up a conversation with a dynamic, successful businesswoman. Which of the following are you most likely to hear her say during the course of your conversation?

Answer choices:
a. "I really don't know how to read very well."
b. "I can't write a grammatically correct sentence."
c. "I'm awful at dealing with people."
d. "I've never been able to think logically."
e. "I'm bad at math."

We all know that the answer is E, because we've heard it so many times. Not just from businesswomen and businessmen, but from actors and athletes, construction workers and sales clerks, and sometimes even teachers and CEOs. Somehow, we have come to live in a society in which many otherwise successful people not only have a problem with mathematics but are unafraid to admit it. In fact, it's sometimes stated almost as a point of pride, with little hint of embarrassment.

It doesn't take a lot of thought to realize that this creates major problems. Mathematics underlies nearly everything in modern society, from the daily financial decisions that all of us must make to the way in which we understand and approach global issues of the economy, politics, and science. We cannot possibly hope to act wisely if we don't have the ability to think critically about mathematical ideas.

This fact takes us immediately to one of the main themes of this book. Look again at our opening multiple-choice question. It would be difficult to imagine the successful businesswoman admitting to any of choices A through D, even if they were true, because all would be considered marks of ignorance and shame. I hope to convince you that choice E should be equally unacceptable. Through numerous examples, I will show you ways in which being "bad at math" is exacting a high toll on individuals, on our nation, and on our world. Along the way, I'll try to offer insights into how we can learn to make better decisions about mathematically based issues. I hope the book will thereby be of use to everyone, but it's especially directed at those of you who might currently think of yourselves as "bad at math." With luck, by the time you finish reading, you'll have a very different perspective both on the importance of mathematics and on your own ability to understand it.

Of course, I can't turn you into a mathematician in a couple hundred pages, and a quick scan of the book should relieve you of any fear that I'm expecting you to repeat the kinds of equation solving that you may remember from past math classes. Instead, this book contains a type of math that you actually *need* for life in the modern world, but which you probably were never taught before.

Best of all, this is a type of mathematics that anyone can learn. You don't have to be a whiz at calculations, or know how to solve calculus equations. You don't need to remember the quadratic formula, or most of the other facts that you were expected to memorize in high school algebra. All you need to do is open your mind to new ways of thinking that will enable you to reason as clearly with numbers and ideas of mathematics as you do without them.

The Math Recession

For our first example, let's consider the recent Great Recession, which left millions of people unemployed, stripped millions of others of much of their life savings, and pushed the global financial system so close to collapse that governments came in with hundreds of billions of dollars in bailout funds. The clear trigger for the recession was the popping of the real estate bubble, which ignited a mortgage crisis. But what created the bubble that popped? I believe a large part of the answer can be traced to poor mathematical thinking.

Take a look at Figure 1, which shows one way of looking at home prices during the past few decades. The bump starting in 2001 represents the housing price bubble. Let's use some quantitative reasoning to see why it should have been obvious that the bubble was not sustainable.

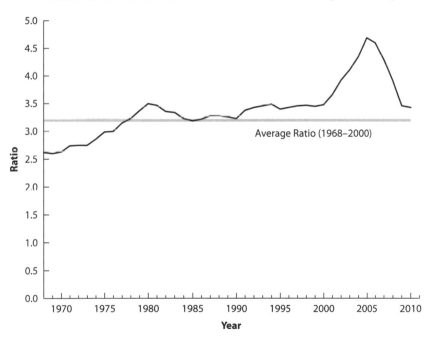

Figure 1. Data used with permission of the Joint Center for Housing Studies of Harvard University. All rights reserved.

Here's how to think about it. As its title indicates, the graph shows the ratio of the average (median) home price to the average income. For example, if the average household income were $50,000 per year, then a ratio of 3.0 would mean that the average home price was three times the average income, or $150,000. The graph shows that the average ratio for the three decades prior to the start of the bubble was actually about 3.2, which means someone with an income of $50,000 typically purchased a house costing about $160,000 (which you find by multiplying 3.2 by $50,000).

Now look at what happened during the housing bubble. After increasing modestly in the 1990s, the ratio began shooting upward in 2001, reaching a peak of about 4.7 in 2005. This was nearly a 50% increase from the historical average of 3.2, which means that relative to income, the average home was about 50% more expensive in 2005 than it was before the bubble. In other words, a family that previously would have bought a house costing $160,000 was instead buying one that cost nearly $240,000.

With homes so much more expensive relative to income, families had to spend a higher percentage of their income on them. In general, a family can spend a higher percentage of its income on housing only if some combination of the following three things happens: (1) its income increases; (2) it cuts expenses in other areas; or (3) it borrows more money. Other statistics showed clearly that average income was *not* rising significantly, and that while homeowners gained some benefit from relatively low mortgage interest rates, overall consumer spending actually increased. We are therefore left with the third possibility: that the housing bubble was fueled primarily by borrowing. With little prospect that incomes would rise dramatically in the future, it was inevitable that this borrowing would be unaffordable and that loan defaults and foreclosures would follow. The only way to restore equilibrium to the system was for home prices to fall dramatically.

Lest you think that this is a case of hindsight being 20/20, keep in mind that these kinds of data were available throughout the growth of the bubble. Anyone willing to think about it should therefore have known that the bubble would inevitably pop, and, indeed, you can find many articles from the time that pointed out this obvious fact. So how did everyone else manage to miss it?

Although it's tempting to blame the problem on a failure of "the system," it was ultimately the result of millions of individual decisions, most of which involved a real estate agent arguing that prices could only go up, a mortgage broker offering an unaffordable loan, and a customer buying into the real estate hype while ignoring the fact that the mortgage payments would become outsized relative to his or her income. In short, many of us ignored the mathematical reality staring us in the face.

That is why I think of the Great Recession as a "math recession": It was caused by the fact that too many of us were unwilling or unable to think mathematically. Perhaps I'm overly idealistic, but I believe that with better math education—and especially with more emphasis on quantitative reasoning—many more people would have questioned the bubble before it got out of hand. We can't change the past, but I hope this lesson will convince you that we all need to get over being "bad at math."

Fear and Loathing of Mathematics

If we as a society (or you as an individual) are going to overcome the problems caused by being "bad at math," a first step is understanding why this form of ignorance has become socially acceptable. This social acceptance is not as natural as it might seem, and in fact is relatively rare outside the United States. Research has shown that infants have innate mathematical capabilities, and it's difficult to find kindergartners who don't get a thrill out of seeing how high they can count; both facts suggest that most of us are born with an affinity for mathematics. Even many adults who proclaim they are "bad at math" must once have been quite good at it. After all, the successful businesswoman of our multiple-choice question probably could not have gotten where she is without decent grades.

My own attempt to understand the origins of the social acceptance of "bad at math" began with surveys of students who took a course in quantitative reasoning that I developed and taught at the University of Colorado. This course was designed specifically for students who did not plan to take any other mathematics courses in college, and the only reason they took this one was because they needed it to fulfill a graduation requirement. In other words, it was filled with students who had already decided that math wasn't for them. When asked why, the students divided themselves roughly into two groups, which I call *math phobics* and *math loathers*. The math phobics generally did poorly in their high school mathematics classes and therefore came to fear the subject. The math loathers actually did pretty well in high school math but still ended up hating it.[1]

Probing further, I asked students to try to recall where their fear or loathing of mathematics may have originated. Interestingly, the most common responses traced these attitudes to one or a few particular experiences in elementary or secondary school. Many of the students said they had liked mathematics until one adult, often a teacher but sometimes a parent or a family friend, did something that turned them off, such as telling the student that he or she was no good at math, or laughing at the student for an incorrect solution. Dismayingly, women were far more likely to report such experiences than men. Apparently, it is still quite common for girls as young as elementary age to be told that, just because they are girls, they can't be any good at math.

Who would say such things to young children, thereby afflicting them with a lifelong fear or loathing of mathematics? Certainly, there are cases where the offending adult is a math teacher with some sort of superiority complex. But more commonly, it appears that the adults who turn kids off from mathematics are those who are themselves afflicted with the "bad at math" syndrome. Like an infectious disease, "bad at math" can be transmitted from one person to another, and from one generation to the next. Its social acceptance has come about only because the disease is so common.

Caricatures of Math

My students taught me another interesting lesson: While they professed fear and loathing of mathematics, they didn't really know what math is all about. Most of their fears were directed at a caricature of mathematics, though admittedly one that is often reinforced in schools.

The students saw mathematics as little more than a bunch of numbers and equations, with no room for creativity. Moreover, they assumed that mathemat-

---

1. Careful readers may recognize that the math loathers wouldn't necessarily say that they are "bad at math," since they had done well at it. However, remember that I was surveying attitudes of college students. Because the math loathers tend to stay just as far away from math as the math phobics, over time they tend to forget the mathematics that they once learned, and then become fearful of confronting it again.

ics had virtually no relevance to their lives, since they didn't plan to be scientists or engineers. It's worth a moment to consider the flaws in these caricatures.

Numbers and equations are certainly important to mathematics, but they are no more the essence of mathematics than paints and paintbrushes are the essence of art. You can see its true essence by looking to the origin of the word *mathematics* itself, which derives from a Greek term meaning "inclined to learn." In other words, mathematics is simply a way of learning about the world around us. It so happens that numbers and equations are very useful to this effort, but we should be careful not to confuse the tools with the outcomes.

Once we see that mathematics is a way of learning about the world, it should be immediately clear that it is a highly creative effort, and that while equations may offer exact solutions, the same may not be true of the mathematical essence. Consider this example: Suppose you deposit $100 into a bank account that offers a simple annual interest rate of 3%. How much will you have at the end of one year?

Because 3% of $100 is $3, the "obvious" answer is that you'll have $103 at the end of a year. This is probably also the answer that would have gotten full credit in your past math classes. But, of course, it's only true if a whole range of unstated assumptions holds. For example, you have to assume that the bank doesn't fail and doesn't change its interest rate, and that you don't find yourself in need of the money for early withdrawal. In the real world, these assumptions are the parts that require far more thought and study —more *real* mathematics— than the simple percentage calculation.

As to my students' assumption that mathematics had no relevance to their lives, our housing bubble example should already show that this is far from the truth. Today, mathematics is crucial to almost everything we do. We are regularly faced with financial choices that can make anyone's head spin; just consider the multitude of cell phone plans you have to select from, the many options you have for education and retirement savings, and the implications of how you deal with medical insurance for both your bank account and your health. Looking beyond finance, we are confronted almost daily with decisions that we can make thoughtfully only if we understand basic principles of statistics, which is another important part of mathematics. For example, your personal decision on whether to use a cell phone while driving should surely be informed by the statistical research into its dangers, and hardly a day goes by without someone telling you why you need this or that to make you healthier or happier—claims that you ought to be able to evaluate based on the quality of the statistical evidence backing them up.

The issues go even deeper when we look at the choices we face as voting citizens. We're constantly bombarded by competing claims about the impacts of proposed tax policies or government programs; how can you vote intelligently if you don't understand the nature of the economic models used to make those claims, or if you don't really understand the true meaning of billions and trillions of dollars? And take the issue of global warming: On one side, you're told that it is an issue upon which our very survival may depend, and on the other side that it is an elaborate hoax. Given that global warming is studied by researchers almost entirely through statistical data and mathematical models, how can you decide whom to believe if you don't have some understanding of those mathematical ideas yourself?

Getting Good at Math

If you have suffered in the past from fear or loathing of mathematics, then I may be making you nervous. Although you may now accept that mathematics is important to your life, a book about math can still seem scary. But it shouldn't. A simple analogy should help.

Just as you don't have to be the Beatles to understand their music, you don't have to be a mathematician to understand the way mathematics affects our lives. That is why you won't see a lot of equations in this book: The equations in mathematics are like the notes in music. If you want to be a songwriter, you'll need to learn the notes, and if you want to be a mathematician (or a scientist or engineer or economist), you'll need to learn the equations. But for the kinds of mathematics that we all encounter every day—the "math for life" that we'll discuss in this book—all you need are those things that we talked about before: an open mind and a willingness to learn to think in new ways.

In fact, I'll go so far as to make you the same promise that I've made to my students in the past. If you read the whole book, and think carefully as you do so, I promise that you'll find not only that you *can* understand the mathematics contained here, but that you'll find the topics both useful and fun.

I have just one favor to ask in return: Help in the cause of battling an infectious disease that has been crippling our society by promising that you'll never again take pride in being "bad at math," and that you'll do what you can to help others realize that being bad at math should be considered no less a flaw than being bad at reading, writing, or thinking.

Crucial Ideas You Didn't Learn in School

Before we delve into all the fun parts, there's one more bit of background we should discuss: why you haven't learned all this stuff previously.

Consider again the housing bubble example. It is clearly mathematical; its analysis requires a variety of different mathematical concepts, including ratios, percentages, mortgages (which use what mathematicians call *exponential functions*), statistics, and graphing. Its practical nature is also clear, since it affected people's lives all over the world. But now ask yourself: Where in the standard mathematics curriculum do we teach students how to deal with such issues?

With rare exceptions (such as college courses in quantitative reasoning), the answer is *nowhere*. The standard mathematics curriculum begins in grade school with basic arithmetic, then moves on in middle and high school to courses in algebra, geometry, and pre-calculus or calculus. In college, you're either in calculus (or beyond) or taking "college versions" of the courses that you didn't fully absorb in high school, such as college algebra. This standard curriculum covers the crucial mathematical skills needed for students who aspire to careers in science, engineering, economics, or other disciplines that require advanced mathematical computation. But it almost completely neglects the kind of mathematics that would be most useful to everyone else, including most of the mathematics that arose in our housing bubble study. Notice, for example, that statistics is not part of the standard curriculum, which means students are not generally taught how to interpret the types of data we discussed in the housing bubble case, or how to analyze graphs like that in Figure 1. And while standard courses may cover exponential functions and the calculations that underlie mortgage payments, they rarely spend any time examining the implications of those payments, or the factors that should go into deciding whether the payments are affordable.

In other words, despite its clear importance, our schools have by and large neglected to teach "math for life." But why? The full answer is fairly complex, but the gist of it lies in the perceived purpose of teaching mathematics. In decades past, mathematics was seen almost exclusively as a tool for science and engineering, so the curriculum was developed with the goal of putting more people on the science and engineering track. This is a good goal (and one that I strongly support), because there's no question that we need many more scientists and engineers. It's also an important goal for a society that strives for equality, because studies show that many of the best-paying and most satisfying jobs are ones that require proficiency with tools like those of algebra and calculus. For this reason, I personally believe that we owe it to children to help keep all their options open while they are under our guidance, and I therefore think that everyone should be required to learn algebra in high school, and ideally to learn calculus as well. (For anyone who doubts that this is possible, I urge you to watch the movie *Stand and Deliver*.)

But as I have already pointed out, this algebra-track learning is no longer enough. The complex decisions we face today require much greater sophistication with mathematical ideas than was the case in the past, and even people who got As in algebra and calculus may not be prepared to evaluate the types of issues that we'll discuss in this book. I'm far from alone in pointing out this need for greater emphasis on quantitative reasoning; many professional societies, including the Mathematical Association of America, have produced reports urging such emphasis. Unfortunately, this type of educational change takes time, and for the most part our high schools and colleges have not yet come around to teaching the kinds of ideas you'll find in this book. In writing it, one of my greatest hopes is that I might make a small contribution to pushing the needed changes along.

2

Thinking with Numbers

A billion here, a billion there; pretty soon you're talking real money.
— **Attributed to Senator Everett Dirksen**

And now for some temperatures around the nation: 58, 72, 85, 49, 77.
— **George Carlin, comedian**

Question: The following statement appeared in a front-page article in the *New York Times*: "[The percentage of smokers among] eighth graders is up 44 percent, to 10.4 percent." What can you conclude from this statement?

Answer choices:
a. The last time this was studied, the percentage of smokers among eighth graders was negative.
b. The last time this was studied, the percentage of smokers among eighth graders was 10.4%, but now it is 44%.
c. The last time this was studied, the percentage of smokers among eighth graders was about 7.2%.
d. There must be a typo, and the first number should have been 4.4, not 44.
e. The author does not understand percentages, because what is written is impossible.

Before I tell you the correct answer, let me tell you a story that I heard a few years ago at a meeting on college mathematics teaching. A group of mathematics faculty had gone to their dean to seek approval for a new course in quantitative reasoning. To explain what the course would cover, they showed him a copy of the textbook they hoped to use (of which I am the lead author). The dean scanned the table of contents, saw that it has a section on uses and abuses of percentages, and immediately said that they could not teach the course because "percentages are remedial, and we don't give college credit for remedial courses." The faculty then turned to the page that contained the quote from the above multiple-choice question and asked the dean to interpret it. Stumped, he soon acceded to the faculty's request for the new course.

The lesson here is that being able to compute numbers is not the same thing as being able to *think* with them. By fifth or sixth grade, most kids have been taught that *percent* means "divided by 100," so we'd certainly expect college students to know that 44% is the same as 44/100, or that 10.4% is the same as 0.104. But interpreting a statement like "up 44 percent, to 10.4 percent" requires thinking at a much higher level. You not only need to understand the meaning of the individual percentages, you also need to think about how they link together. In this case, we're looking for a number that, if you increase it by 44%, ends up at 10.4%. The correct answer is therefore C, because 10.4% is 44% higher than 7.2%. (You can check this answer as follows: If you start from 7.2%, then an increase of 44% means an increase of 0.44 × 7.2%, which is approximately 3.2%. Adding 3.2% to the starting value of 7.2% gives you the 10.4% result.)

Statements like "up 44 percent, to 10.4 percent" appear often in news reports, and once you understand them, you can see that they are a perfectly reasonable way of conveying information. But as the college dean story shows, even many well-educated people were never taught how to interpret them. There are at least two reasons why standard curricula do not cover such skills. First, they don't fit in well with the traditional progression of mathematics. The idea that 44% is 44/100 is nothing more than division, and therefore can be taught to students in elementary school. In contrast, the interpretation of "up 44 percent, to 10.4 percent" requires an implicit understanding of algebra (because finding the starting point of 7.2% involves the process of solving for an unknown variable), along with abstract reasoning skills that most students don't acquire until at least high school.

A second reason that these skills are rarely taught is that they are more difficult to teach. For example, while "percent" *always* means "divided by 100," the percentage statements in news reports are varied and complex, and sometimes not even stated correctly. There is no single formula that will always work for interpreting such statements, so we generally learn to deal with them through practice and experience.

In the rest of this chapter, I'll present examples designed to give you some experience at thinking with the kinds of numbers we see regularly in the news. They should be fun in and of themselves, but I've chosen them primarily to help you build a basic skill set for quantitative reasoning that we'll then be able to use in later chapters, in which we'll focus our attention on some of the major issues of our time.

Thinking Big

Most everyone knows that ten is ten times as much as one, that one hundred is ten times as much as ten, and that one thousand is ten times as much as one hundred. Knowing those, the meanings of "ten thousand" and "one hundred thousand" are fairly obvious. But beyond that, relatively few people realize that you have to multiply by *one thousand* to make each jump from million to billion to trillion, and even fewer have an intuitive understanding of what these jumps really mean. I don't think it's an exaggeration to say that, for most people, the differences between million, billion, and trillion are primarily in their first letters. Given how often we hear such numbers in the news, it's clearly important to build better intuition for large numbers. Let's do that by discussing a few simple examples.

Million-dollar athlete. Imagine that you are an elite athlete and sign a contract that pays you $1 million per year. How long would it take you to earn your first *billion* dollars? If you remember that a billion is the same thing as a thousand million, then the answer is obvious: It would take one thousand years to earn $1 billion at a rate of $1 million per year. But obvious as the numbers may be, it takes some thought to get this result to sink in. A salary of $1 million per year would strike most people as almost unimaginable riches, yet it would take a *thousand years* of such a salary to earn your first billion—which still wouldn't put you on the Forbes 400 list of the world's richest people.

Hundred-million-dollar CEO. Now assume you're on the board of directors of a large corporation, and there's a proposal on the table to offer the CEO a pay package worth some $100 million per year (an amount that is high but not unheard of during recent years). The company is profitable and the CEO is a smart guy, so you're thinking you'll vote in favor. But then you wonder: Are there other ways the company could spend the same money that might produce greater long-term value for shareholders? It's a subjective question, of course, but here's a thought: Average salaries for research scientists (with PhDs in subjects such as physics, chemistry, and biology) are around $80,000 to $100,000 per year. Let's suppose your company is willing to pay on the high end *and* to add another $100,000 for lab equipment and other research expenses. Then you'd need $200,000 for each scientist you hire, which means that the $100 million that you were going to pay to the CEO could alternatively be used to hire 500 research scientists (because $100 million ÷ $200,000 = 500). I know that some CEOs are very talented people, but you're going to have a hard time convincing me that any one person could produce the same long-term value to your company that you'd get from having 500 additional scientists working full-time to help your company come up with new inventions and products. And if you really want to think long term, let's allow each of those 500 scientists to take one day a week to go help out with science teaching at a local school. If we assume that each scientist spends the day with a group of 30 kids, that's 15,000 students who will be touched by these weekly visits. Aside from the general good that would come of this, don't forget that all of them are potential future customers—or future employees—who may remember that *your* company provided the opportunity.

A billion here, a billion there. Now let's move into the realm of the "real money" alluded to in the famous aphorism that opens this chapter. The same math that shows that $100 million could hire 500 scientists means that $1 billion could hire 5,000 of them. Going a step further, the $23 billion that Goldman Sachs initially set aside for its bonus pool in 2009 (they later reduced it to about $16 billion) would allow the hiring of more than 100,000 scientists. Even if you change the assumption from $200,000 to $2 million per scientist, thereby allowing plenty of money for building construction, staff expenses, and higher salaries, you could still hire more than 10,000 scientists. In other words, if the $23 billion were sustainable year after year, "Goldman Scientific" could become the largest single research institution in the world, with an annual operating budget roughly ten times that of major research institutions such as MIT or the University of Texas at Austin. Since I'm a fan of human space exploration, I'll also point out that $23 billion is about 25% larger than NASA's budget (roughly $18 billion in 2011), which means it is somewhat more than a presidential commission said would have been needed to keep NASA's cancelled "return to the Moon" program on track. So it seems to me that Goldman missed an opportunity to be on the forefront of future business opportunities in space, opportunities likely to offer far more long-term benefit for shareholders than lavishing large paychecks on wizards of finance.

Government money. Even Goldman pales in comparison to the sums that we regularly hear about with government programs. The biggest sum that's regularly in the news is the federal debt, for which you might want to calculate your share. If you divide the roughly $15 trillion debt (late 2011) by the roughly 310 million people in the United States, you'll find that each person's share of the debt is close to $50,000, which means that an average family of four owes close to $200,000 to future generations—significantly more than it owes for its home. And at the risk of really depressing you, I'll remind you that the debt is not only a burden on the future, but also a burden today because the government must pay interest on it. In 2010, for example, the interest totaled $414 billion[2]—which is *more than the total spent by the federal government on education, transportation, and scientific research combined*. Worse, the only reason the interest payment was so "low" was because of record low interest rates. If interest rates rise back up to something more like their average for recent years, the annual interest payments on the current debt could easily double or triple, and that's before we even consider the fact that the debt is still rising. Perhaps, as some politicians argue, we've had no choice but to borrow (and continue to borrow) so much money. But when you consider

---

2. Most news reports said the 2010 interest was "only" $196 billion, which is the interest that the government paid on debt held by the public (and by other nations). The $414 billion "gross interest" includes interest on government accounts such as the Social Security trust fund, which makes it a better measure of how interest affects the government's present and future obligations.

what else we might do with the money going to interest alone, it sure makes you think that there ought to be a better way.

Counting stars. Let's turn to some big numbers that are less depressing and more amazing. One of my favorites is the number of stars. As you probably know, our Sun is just one of a great many stars that, together, make up what we call our Milky Way Galaxy. The galaxy is so big that no one knows its exact number of stars, but estimates put the number at a few hundred billion. To make the arithmetic easier, let's just call it "more than 100 billion." Now, suppose that you're having trouble going to sleep tonight, so you decide to count stars. How long would it take you to count 100 billion of them? If we assume that you can count at a rate of one per second, then it would take 100 billion seconds.[3] You can then divide by 60 to convert the 100 billion seconds to minutes, divide by 60 again to convert it to hours, divide by 24 to convert it to days, and divide by 365 to convert it to years. Try it on your calculator, and you'll find that 100 billion seconds is almost 3,200 years. In other words, it would take more than 3,000 years just to *count* 100 billion stars in our galaxy, assuming that you never take a break, never go to sleep, and manage to stay alive for a few thousand years. And that's just the stars in *our* galaxy.

If you multiply the 100 billion stars in a typical galaxy by the estimated 100 billion galaxies in the known universe, you'll find that the total number of stars in our universe is about 10,000,000,000,000,000,000,000 (a 1 followed by 22 zeros, or 10^{22}), which you could say as "10 billion trillion," or "10 million quadrillion," or "10,000 billion billion." But rather than giving it a name, I prefer a more interesting comparison. You can estimate the number of grains of sand in a box by dividing the volume of the box (which is its length times its width times its depth) by the average volume of a single sand grain. In the same basic way, you can estimate the number of grains of sand on all the beaches on Earth by finding the total volume of beach sand and dividing by the average volume of sand grains. Estimating the total volume of beach sand

---

3. A counting rate of one per second may sound pretty easy when you think of the starting numbers like one, two, and three. But it gets more difficult when you reach numbers like, say, "thirty-seven billion, four hundred ninety-two million, six hundred eighteen thousand, two hundred forty-four"—and then have to immediately remember what comes next.

on Earth is not as difficult as it sounds, though like most measurements, it's much easier if you use metric units. A quick Web search will tell you that the total length of sandy beach on Earth is about 360,000 kilometers (about 220,000 miles), and the average beach is about 50 meters wide and 4 meters deep. I'll leave the rest of the multiplication (and division by the average sand grain volume) to interested readers, and just tell you the amazing result: *the number of grains of sand on all the beaches on Earth is comparable to the number of stars in the known universe.* Next time you're thinking about whether there might be other civilizations out there, remember that in comparison to all the stars in the universe, our Sun is like just one grain of sand among all the grains on all the beaches on Earth combined.

Until the Sun dies. As the examples of star counting show, astronomy is a subject full of amazement, and one that should make us proud to be members of a species that has managed to learn such incredible things about our universe. But astronomy sometimes seems scary, too, especially when you learn, for example, that the Sun is doomed to die. Fortunately, a little math should relieve any concerns you might have. The Sun is indeed doomed to die, but not for about 5 billion years. How long is 5 billion years? One way to put it in perspective is to compare it to a human lifetime. If we assume a lifetime of 100 years, then 5 billion years is about 50 million lifetimes. It turns out that 100 years also happens to be close to 50 million minutes (which you can see by taking 100 years and multiplying by 365 days in a year, 24 hours per day, and 60 minutes per hour). We can therefore say that a human lifetime compared to the remaining life of the Sun is like a mere minute in a long human life. Human creations register only a little more on the Sun's time scale. The Egyptian pyramids have often been described as "eternal," but at their current rates of erosion, they will have turned to dust within about 500,000 years. That may sound like a long time, but the Sun's remaining lifetime is some 10,000 times longer. Clearly, we have more pressing things to worry about than the eventual death of our Sun.

Another way to consider the Sun's remaining 5 billion years is to think about what would happen if we ended up doing ourselves in. No matter how much damage we do to our planet, we won't wipe out life entirely. If we cause our own extinction, it's likely that some other species will eventually evolve intelligence as great as ours, giving Earth another chance to have a civilization that makes good rather than destroying itself. There's no way to know exactly how long it would take for the next intelligence to arise, but I'd say that 50 million years is a pretty conservative guess. In that case, if the intelligent beings that rise up 50 million years from now also wipe themselves out, another intelligence could presumably emerge some 50 million years after that, and so on. You might think that at 50 million years per shot, Earth would quickly run out of opportunities. But it wouldn't: The 5 billion years remaining in the Sun's lifetime would be enough for Earth to have 100 more chances for an intelligent species to rise up, each 50 million years after the last one. It's truly incredible to think about, and it makes you think that, eventually, there would be a species smart enough to travel to the stars and thereby eliminate worry about what happens when the Sun dies. We can only hope that species will be us.

Lunch with your students. Having talked about numbers in the millions, billions, and trillions, it's easy to start thinking that anything in the thousands must be small. But even those numbers are much larger than we usually recognize. Imagine that a university with 25,000 students (typical of many state universities) hires a new president. Thinking that he should get to know the students, the president offers to meet for lunch with groups of 5 students at a time. If all 25,000 students accept, how long will it take the president to finish all the lunches? Again, the basic math is straightforward. If he holds the lunches 5 days a week, with 5 students at a time, then he'll be dining with 25 students per week. If we leave 2 weeks off for the winter holidays and 10 weeks for summer, he could have these lunches 40 weeks per year, which means the lunches would include a total of $40 \times 25 = 1,000$ students each year. At that rate, it would take him 25 years to get through the lunches with all 25,000 students—but, of course, that wouldn't work, since most of the students would have graduated long before getting their turn.

Incidentally, similar thinking probably explains why "special interests" have become so dominant in politics. The U.S. House of Representatives has 435 members; dividing this number into the U.S. population of about 310 million people, we find that each representative has an average of more than 700,000 constituents. If we assume a 40-hour workweek, 50 weeks per year, then each representative has about 2,000 working hours per year, or 4,000 hours during a two-year term. If you divide that by the 700,000 constituents, you'll find that a representative could at best devote about 20 seconds to each constituent (on average). Given this reality, along with the reality that it can take millions of dollars to run a campaign, it's no wonder that the representatives devote most of their listening time to the relatively small numbers of people who fund the bulk of their campaigns.

Stadium lottery. A different type of big-number thinking requires putting various odds into perspective. As an example, imagine watching a football game in a stadium filled to capacity with 50,000 people. The announcer comes on and says that if everyone is willing to ante up $500 each, the league will pick one person at random to receive a multimillion-dollar prize. Would you pay the $500? Probably not; after all, when you look around at a stadium full of people, it seems almost impossible to believe that you'd be the one person selected at random, and it certainly wouldn't seem worth spending $500 for that tiny chance. Yet outside the stadium, nearly half of all Americans play this very game every year. That's because people who play the lottery (which about half of all Americans do) spend an average of about $500 per year on their lottery tickets, while each person's chance of being a big winner is no bigger than the chance of being that one person selected in the stadium. In fact, it's actually smaller, since the trend has been for lotteries to offer larger prizes with worse odds. To put it a different way, even if you spend $500 per year—which adds up to $20,000 over a 40-year playing "career"—the chance that you'll ever be one of the big winners is only about 1 in 50,000. So to all the lottery players out there, consider this statement of fact: While *someone* will surely win, I can be 99.998% certain that it won't ever be *you*.[4] Still want to play, or can you think of better uses for your $20,000? As a widely circulated Internet message says, the lottery is essentially "a tax on people who are bad at math."

The same basic ideas apply to gambling of all types. When you walk into a casino, the odds have been stacked against you—that's why the casino has money to offer you all those free drinks and other enticements. If you think of yourself in the stadium full of people, you'll probably realize how crazy it is to start gambling. But when it's just you and the machine, or you and the card dealer, it can suddenly seem like you must be bound to win. Moreover, the gambling companies have spent hundreds of millions of dollars on research

---

4. Because the 1 in 50,000 chance of winning means that your chance of *not* winning is 49,999/50,000, which is 0.99998, or 99.998%.

to find the best ways to convince you to keep playing, with lighting, bells, and other tricks of the trade designed to make you think you have more of a chance than you really do. Frankly, I think this gives the casinos a fundamentally unfair advantage over their patrons, and if it were up to me I'd require all casinos to post large warning labels, much like those we require on cigarettes. In this case, they could read something like: "WARNING: The games in this facility are set up so that the odds are stacked against you. While an individual may occasionally come out ahead after any particular play, continued playing virtually guarantees that you will lose money in the end."

Dealing with Uncertainty

In math classes, you were probably told to assume that the numbers you dealt with were always exact. In science classes, you may have learned that measurements have associated uncertainties, and learned techniques for dealing with those uncertainties. The situation is more difficult in the real world, where we may not even have a good way to estimate the uncertainty associated with the numbers we encounter.

Consider the forecasts we hear each year about future budget deficits. As recently as 2008, for example, the president's budget office predicted that the deficit for 2009 would be $187.166 billion. Notice that the number was stated to the nearest $0.001 billion, which is the same as the nearest $1 million. When 2009 ended, the actual deficit turned out to be $1.42 *trillion*— which means that although the deficit prediction had been stated as though we knew it to the nearest million dollars, in reality we didn't even know it to the nearest *trillion* dollars!

In fairness, the budget office is staffed by pretty smart people, and they were well aware that they couldn't really know the future deficit to the nearest million dollars. Their full report included hundreds of pages that outlined various assumptions that would have had to be true for the numbers to come out exactly as predicted, along with descriptions of various uncertainties that could also affect the predictions. However, when budget numbers appear in the news media, all those caveats usually disappear, which can mislead you into thinking that the numbers are known far better than they really are.

The fact that numbers are so often reported without clear descriptions of their uncertainties means we must develop ways of looking critically at all the numbers we encounter. Rather than proceeding through specific examples as we did with big numbers, I'll suggest four general ways of thinking about uncertainties.

Accuracy versus precision. Although many people interchange the words *accuracy* and *precision*, they are not quite the same thing. To understand the distinction, imagine that you actually weigh 125.2 pounds, and that you check your weight on two different scales. One scale is the old-fashioned type that you can at best read to about the nearest pound, and it says you weigh 125 pounds. The other scale is digital, and it says you weigh 121.44 pounds. We say that the reading on the digital scale is "precise to the nearest 0.01 pound," while the reading on the old-fashioned scale is "precise to the nearest pound." This means the digital scale is more precise. However, because the old-fashioned scale got closer to your actual weight, it is more accurate. In other words, accuracy describes how closely the measurement approximates the true value, while precision describes the amount of detail in the measurement.

You can probably see how unwarranted precision can cause problems. For example, stating a weight as 125 pounds implies that you know it to the nearest pound, while stating a weight as 121.44 pounds implies that you know it to the nearest 0.01 pound. In this case, the fact that your actual weight was 125.2 pounds means the first statement was true (a weight of 125 really is correct to the nearest pound) while the second statement was false. More generally, stating a number with more precision than is justified is always deceptive, because it implies that you know more than you really do.

Let's apply this idea to the budget deficit example. When the 2009 deficit projection was stated to the nearest $1 million, it implied that it was accurate within this amount. Given that the projection turned out to be wrong by more than $1 trillion, and that a trillion is a million times a million, the actual uncertainty in the budget estimate was *a million times worse* than the implied uncertainty of $1 million. We can't really blame the budget office, since they had those hundreds of pages that explained all the caveats. The blame, if any, should go to the media that reported the number as though we really did know it that well.

The recent census provides another good example. According to the published reports, the census found that the U.S. population on April 1, 2010, was 308,745,538. But there's no way that anyone could really know the population exactly. Aside from the inevitable difficulties of counting, the fact that an average of about eight births and four deaths occur *each minute* in the United States means that you could only know the exact population if there were some way to count everyone instantaneously, while the census was carried out over a period of many months. Like the budget office, the Census Bureau was well aware that the number was not really known as well as its precision implied. In fact, if you read the full census report, you'll find that the Census Bureau estimated the uncertainty in the population count to be at least three million people, meaning the actual population could easily have been three million higher or lower than the reported value.

The bottom line is that many of the numbers that we hear in the news are reported with more precision than they deserve, falsely implying a level of accuracy that doesn't really exist. So the first lesson in dealing with uncertainty in the news is to beware of any number you hear, and to think carefully about whether it can really be as precise as reported. Given the news media's propensity to leave out all the important caveats, when possible you should go back to original sources (such as the budget documents or Census Bureau reports) to find out what has been ignored.

Random versus systematic errors. Numbers may be inaccurate for a variety of different reasons, but in most cases we can divide those reasons into two broad classes: *random errors* that occur because of unpredictable events in the measurement process, and *systematic errors* that result from some problem in the way the measurement system is designed.

Consider the potential sources of inaccuracy in the census count of the U.S. population. Some errors may occur because people fill out the census surveys incorrectly, or because census workers make mistakes when they enter the survey data into their computers. These types of accidental errors are random errors, because we cannot predict whether any individual error overcounts or undercounts the population. In contrast, consider errors that occur because census workers can't find all the homeless or all of the very poor, or because undocumented aliens try to hide their presence. These are systematic errors that arise because the system is unable to account for all the people in those groups, and these particular systematic errors can only lead to an undercount. Other types of systematic errors can lead to overcounts; for example, college students may be counted both by their parents and in their housing at school, and children of divorced parents may be counted in both households.

Perhaps the most important distinction between random and systematic errors is that while there's nothing you can do about random errors after they've occurred (though well-designed systems can minimize the likelihood of their occurrence), you can correct for systematic errors if you are aware of them. For example, by looking for the homeless, the poor, or undocumented aliens with extra care in a few selected areas, the Census Bureau can estimate the amount by which its standard processes tend to undercount these groups. Indeed, the Census Bureau has data available that should in principle allow it to make its population estimate more accurate—but it is allowed to use these data only for limited purposes. Part of the problem revolves around a constitutional question: The U.S. Constitution (Article 1, Section 2, Subsection 2) calls for an "actual enumeration" of the population. Those who oppose the use of statistical data to improve the population estimate point out that "enumeration" seems to imply a one-by-one count. Those who favor using the statistical data point out that an exact count is impossible, and therefore focus on the word "actual," arguing that statistics can help us get closer to the actual value. Of course, the real issue is probably more political than constitutional: Democrats tend to favor the use of statistical data because it leads to higher numbers of people who tend to vote Democratic, while Republicans oppose the use of statistical data for the same reason. Note that this debate is not just about voting. The census results affect the makeup of Congress and of state legislatures, because they are used to apportion political representation by state and by locality. The census results also have economic value, because states and cities receive allotments of federal money based on their populations.

Absolute versus relative errors. There are two basic ways to think about the sizes of errors. First, we can think about the *absolute error*, meaning the actual amount by which a given number differs from its true value. Alternatively, we can consider the *relative error*, which describes the size of the error in comparison to the true value. A simple example should illustrate the point. If the government ever managed to predict the budget deficit to within about $1 million, we'd be very impressed, because $1 million is so small compared to the trillions of dollars that the government collects and spends. But if your electric company overcharged you by $1 million, the error would seem enormous. In other words, both cases have the same absolute error of $1 million, but the relative error is much smaller for the deficit than for your electric bill. By the way, in case you haven't thought it through fully yet, this idea explains the famous quote from Senator Dirksen: Politicians can throw around dollars like "a billion here, a billion there" because billions are relatively small in a federal budget that is measured in trillions, but there's no doubt that in absolute terms, we're talking "real money."

Measurements versus models. So far we've talked about the interpretation of numbers and their uncertainties, but it's also important to consider where numbers come from in the first place. For example, a weight on a scale represents a simple measurement, while a prediction about a future budget deficit represents the result of a complex economic model that may have tens of thousands of variables, all evaluated by a computer that performs millions of calculations. Although it's possible that a weight measurement could have a relative error as large as that of a budget prediction, it's also pretty obvious that the budget prediction has many more ways to go wrong. Economists and scientists test models by using them to try to reproduce measurements made in the past. For example, if your economic model can successfully "predict" last year's deficit from information that was available before the year began, then you would have at least some reason to trust its prediction for next year. Of course, unforeseen circumstances could still make the model quite wrong, as was the case with the 2009 deficit prediction that we've discussed. Among other problems, the model used in that prediction did not take into account the collapse of the housing market or the massive government bailouts that followed.

Apples and Oranges

The famous saying that you can't add apples and oranges reflects a deeper idea about the numbers we encounter in daily life, which is that numbers are almost always associated with some type, or *unit*, of measurement. If you have five apples and three oranges, you can think of the units as apples and oranges, and because these units are different, you can't combine them.

Units provide crucial context to numbers. If I say that a person weighs 75, the meaning is quite different if I mean pounds than if I mean kilograms. Similarly, a temperature of 32 is pretty hot if you're in Europe, where temperatures are reported on the Celsius scale, but it's freezing on the Fahrenheit scale used in the United States. Of course, units alone may not provide all the context needed; the George Carlin quote at the beginning of the chapter is funny not because he didn't distinguish between Celsius and Fahrenheit, but because he left out the critical context of locations.

For the most part, news media are pretty good about stating units; you'll rarely hear a number reported without it being clear whether the number represents dollars, pounds, people, or something else. So our reason for discussing units has less to do with the news and more to do with the ways in which they can help us think about quantitative problems. In fact, *unit analysis* is arguably the simplest and most useful of all problem-solving techniques—yet it is rarely discussed in math classes (though often covered in science classes). To get started with unit analysis, you need only remember two simple ideas: the word *per* implies division, while *of* implies multiplication.

As an example, imagine that you're trying to figure out the gas mileage you're getting, but aren't sure how to do it. If you remember that gas mileage is usually given in units of miles per gallon, you'll immediately recognize that you need to take something with units of miles and divide it by something with units of gallons. From there, it's a small step to realize that you should divide the number of miles you've driven since you last filled your gas tank by the number of gallons it takes to fill up. For example, if you drove 200 miles on 8 gallons of gas, then your mileage is 200 miles ÷ 8 gallons = 25 miles per gallon. Similarly, you can always remember that speed is a distance divided by a time just by recalling that we measure highway speeds in "miles per hour."

Cases with *of* are similarly easy. Suppose you buy 10 pounds of apples at a price of \$3 per pound. The word *of* (in "price of \$3") tells us to multiply, so the total price is 10 pounds × \$3/pound = \$30. Notice how the pound units cancel out to leave dollars: This happens because the first number is in pounds, while the second number divides by pounds, and anything divided by itself is just a plain number one.

Unit analysis can be done at more sophisticated levels; in fact, it has led to numerous important scientific insights. For most of the things you'll encounter in daily life, however, the rules with *of* and *per* are all you need to know.

Back to Percentages

We began this chapter with a multiple-choice question demonstrating that although the basic idea of percentages is easy, the uses of percentages can be surprisingly complex. So before we leave our discussion of basic skills for quantitative reasoning, let's look at a few more of the ways that percentages are often used or abused.

"Of" versus "more than." One snowy season in Colorado, a television news reporter stated that the snowpack was "200% more than normal." At the same time, a reporter on another channel said that it was "200% of normal." The two statements sound very similar, but they are actually inconsistent. Here's why: Because *percent* means "divided by 100," 100% means 100 divided by 100, which is just 1; that is, 100% is just a fancy way of saying the number 1. By the same reasoning, 200% means 2, 300% means 3, and so on. Now, suppose the normal snowpack for that time of year was 100 inches. Because *of* means multiplication, 200% *of* normal means "2 times normal," implying a snowpack of 200 inches. The statement 200% *more than* normal must therefore imply a snowpack of 300 inches, because it is 200 inches more than the normal 100 inches. Given the different meanings of the two news reports, you'd probably want to know which one was correct. Unfortunately, without being given the actual snowpack numbers, there's no way to know which reporter used words correctly and which one did not.

The lesson of the snowpack example is that you have to be very careful when listening to statements that use "of" and "more than" (or "less than"), because people often mix them up even though they have different meanings. Just to be sure the point is clear, consider a stock that sells for $10 per share on January 1. If the share price on July 1 is 200% of the price on January 1, it means the price has risen to $20; but if the share price on July 1 is 200% more than the price on January 1, then it means the price has risen to $30. Similarly, if the share price on July 1 is 25% of the price on January 1, it means the price has fallen to 25% of $10, or $2.50; but if the share price on July 1 is 25% less than the price on January 1, then the price has only fallen to $7.50.

The same type of confusion can occur even without percentages, and it is so pervasive that I've even found it done incorrectly in textbooks. For example, the planet Jupiter is about five times as far from the Sun as Earth is, but I've seen books that say Jupiter is "five times *farther* from the Sun" than Earth—which would imply six times as far, not five times as far. Similarly, because the distance of Mars from the Sun is about 1.5 times the distance of Earth from the Sun, the statement "Mars is 1.5 times farther from the Sun than Earth" is false; the correct statement would be that it is 50% (or 0.5 times) farther from the Sun. Again, the lesson is twofold: When speaking (or writing), be careful not to mix up words that imply *of* and those that imply *more than*; when listening (or reading), remember that others may not be so careful, so if it's important, find a way to verify the statement before taking it at face value.

Percentage more or less. Suppose that in a difficult economy, your boss asks you to take a 10% pay cut this year, but promises to give it back in the form of a 10% pay raise next year. Let's see how it works out by imagining that your pay rate is $10 per hour. Because 10% of $10 is $1, the pay cut will lower your hourly rate to $9. Therefore, when you get the 10% pay raise next year, it will mean 10% of $9, which is $0.90 . . . meaning that your raise will take you to $9.90 per hour. In other words, the 10% cut followed by the 10% raise does *not* return you to where you started.

Here's another surprising example: In 2008, the stock market (as measured by the Dow Jones Industrial Average) lost about 34% of its value. Over the next two years (through the end of 2010), the market posted a 32% gain. If you subtracted, you might therefore think that the market at the end of 2010 was only about 2% below where it started in 2008; in fact, the market was still nearly 13% below where it had started in 2008.[5]

How do these strange things happen with percentages? The best way to understand it is with a simpler example. Suppose you are comparing the prices

---

5. In case you want to verify the percentages, here are the data: the DJIA began 2008 at 13,265, ended 2008 at 8,776, and ended 2010 at 11,578.

of a $50,000 Mercedes and a $40,000 Lexus. If we work with the straight numbers, we find symmetry: We can say either that the Mercedes costs $10,000 more than the Lexus, or that the Lexus costs $10,000 less than the Mercedes. But notice what happens when we shift to percentages. Because $10,000 is 25% of $40,000, we would say that the Mercedes costs 25% more than the Lexus. When we go the other way around, however, we are starting with the Mercedes price of $50,000; because $10,000 is only 20% of $50,000, we find that the Lexus costs 20% less than the Mercedes. The general rule to remember is that percentages always depend on the number we use as the reference value. When we use the Lexus price of $40,000 as the reference value, then $10,000 represents 25%; when we use the Mercedes price of $50,000 as the reference value, then the same $10,000 represents only 20%. That is why the statements "The Mercedes costs 25% more than the Lexus" and "The Lexus costs 20% less than the Mercedes" are both true.

It's worth noting that this mathematics of percentages means that it's much easier for numbers to go up by large percentages than to go down. For example, suppose the price of gasoline were $2 per gallon. The price could easily go up by 100%, to $4 per gallon, or even by 200%, to $6 per gallon. But if the price *fell* by 100%, it would mean the gasoline was being given away for free, and it would be impossible for the price to fall by 200%, since that would mean the gas station would have to *pay you* when you filled up.

Percentages of percentages. Look back at our chapter-opening multiple-choice question about the statement "up 44 percent, to 10.4 percent." One of the main reasons this statement can sound confusing is that it involves percentages of percentages; that's why it took some thought to realize that the "up 44 percent" was based on a starting value of 7.2% (so that the 44% increase raised the value to 10.4%).

Statements like "up 44 percent, to 10.4 percent" at least have the benefit of a clear meaning, once you think them through. Unfortunately, other statements with percentages of percentages can be more ambiguous. For example, suppose you learn that your bank charges 8% interest on loans for new cars, then hear the next day that it has raised its rates 1%. Does this mean the new rate is 8.08% (because 1% of 8% is 0.08%), or does it mean that the new rate is 9%? Without further information, there's no way to know.

The only way to avoid this type of ambiguity is to make some general agreement on language. Although I'm unaware of any formal definitions, it has become conventional to use the term *percentage points* to mean something additive, and the % sign to mean something multiplicative. In that case, a rate rise from 8% to 9% would be considered a rise of one percentage point, while a rise of 1% would mean the rise from 8% to 8.08%. Not everyone follows this convention, however, so you must still be very careful when you interpret statements about percentages of percentages.

Percentage of what? During the 2004 presidential campaign, Democratic candidate John Kerry got into a running debate with President George W. Bush over whether the war in Iraq was essentially a "U.S. war" (as Kerry claimed) or whether it was a war being fought by an "international coalition" (as Bush claimed). The debate turned mathematical when each candidate used percentages to back up his claim. Kerry supported his claim by stating that, although other countries had sent troops to Iraq, the United States had 90% of the troops and was bearing 90% of the casualties. Bush countered by stating that, in fact, the United States was supplying only 40% of the troops and bearing only 40% of the casualties. The argument continued for weeks, playing out both in the presidential debates and in the news.

You might wonder how the debate could have gone on so long, since it seems it should have been easy for someone to figure out whether the correct number was 40% or 90%. In fact, both numbers were right; they were just based on different things. Here's how: When Kerry spoke of the international coalition, he meant countries that had sent troops to Iraq, and the United States represented 90% of those troops. But when Bush spoke of the international coalition, he meant *all* troops fighting in Iraq, including the Iraqis themselves. Since the Iraqis had a lot of their own troops involved in the war, the United States represented only 40% of that total.

The lesson from this case is that a percentage is always a percentage of *something*, and percentages are meaningless unless you are very clear about *what* that something is. Here's another example: In June 2010, the monthly government survey showed that the United States lost 261,000 jobs in the previous month and that the unemployment rate fell from 9.6% to 9.5%. Wait, you say: If we *lost* jobs, shouldn't the unemployment rate have gone up, rather than down? Again, the question is, "Percentage of *what?*" If the unemployment rate measured the percentage of people without jobs, then it would have to go up when jobs were lost. But if you go to the Web site of the Department of Labor, you'll find that what the unemployment rate actually measures is more complex; it is essentially the percentage of people unemployed among those who are either working or *actively looking* for a job. In other words, someone who loses hope and stops looking for work is no longer considered unemployed; similarly, a mom or dad who decides to stop working to stay home with the kids is out of a job but not counted as unemployed. So while the drop in the unemployment rate by itself might sound like good news, understanding what's really being measured suggests that it probably dropped because a lot of people without jobs stopped trying to find one.

The importance of knowing *what* when it comes to percentages leads to one last rule for this chapter: *Never try to average percentages.* To see why, imagine a basketball player who hits 80% of his free throws during the first half of the season and 90% during the second half. Though it might be tempting to say that his season average was 85%, you can't do that, because it's unlikely that the *what*—in this case, the number of free throw attempts—was the same in both cases. For example, suppose that he had 10 free throws during the first half of the season and 90 during the second half. The 80% from the first half means he made 8 out of his 10 shots, and the 90% from the second half means he made 81 out of his 90 shots. Therefore, his season total was 89 out of 100 free throws, which is 89%, not the 85% that we'd guess by averaging the percentages.

Math for Life

You've probably noticed that I've started a pattern in which each chapter begins with a question. Now, I'll start a second pattern by ending each chapter with a brief summary that emphasizes how and why I think we can build a society that is better equipped at "math for life."

This chapter covered the importance of four key ideas that apply to almost every use of numbers that you'll encounter in the news or elsewhere: (1) being able to put large numbers in perspective; (2) analyzing the uncertainties associated with most numbers in the real world; (3) understanding the units that accompany numbers and give them context; and (4) being able to interpret the many contexts in which we encounter percentages, and being aware of when those percentages may have been misused.

If you miss out on even one of those four key ideas, you'll inevitably miss out on understanding some of the issues you encounter daily in the news. Indeed, I challenge you to find any newspaper (physical or online) in which all four ideas don't arise somewhere, and usually they'll each arise in multiple stories. So while you may not encounter exactly the same examples that I've offered in this chapter, the type of thinking that goes into them is something you should continue to practice and build. Otherwise, you'll find there's no way to make a rational decision when it comes time to cast your votes, state your opinions, or even decide which media reports to trust. Most important, remember that others who still are "bad at math" may be making their own irrational decisions even as we speak. Help them, please, by spreading the gospel of learning how to think with numbers.

3

Statistical Thinking

Statistical thinking will one day be as necessary for
efficient citizenship as the ability to read and write.
—**H. G. Wells**

With proper treatment, a cold can be cured in a week.
Left to itself, it may linger for seven days.
—**Medical folk saying**

Question: You want to know whether Americans generally support or oppose the new health care plan. Which of the following
approaches is most likely to give you an accurate result?

Answer choices:
a. Short interviews with just 1,000 randomly selected Americans,
conducted by a professional polling organization such as Gallup
or the Pew Research Center
b. In-depth interviews with 10,000 Americans selected at random
from among those who have been hospitalized in the past year,
conducted by a professional polling organization
c. Short interviews with 100,000 Americans, chosen by asking 100
randomly selected people at each of 1,000 grocery stores at 9 a.m.
on a particular Monday morning, conducted by volunteers working for a citizens' group
d. A poll in which more than 2 million people register their opinions online, conducted by a television news channel
e. A special election held nationwide, in which all registered voters
have an opportunity to answer a question about their opinion on
the health care plan

I know some of you are thinking that this one is too easy. But before you jump to that conclusion, try doing your own poll by giving this multiple-choice question to a group of friends or family members, or a group of high school or college students, or a set of your business associates. I'd be willing to bet that a lot of people will choose a wrong answer.

Take answer D, for example. The fact that these polls are commonly offered by news shows, and that millions of people participate in them, seems to suggest that a lot of people put some stock in them. But all such polls inevitably suffer from at least two major biases. First, only people who watch a particular newscast will know about the poll, and these days the different networks (consider Fox and MSNBC) tend to have viewers with very different political views. Second, people with strong feelings on the issue are much more likely to take the time to go online for the poll than those who are more middle-of-the-road, and there's usually nothing to prevent those with the strongest feelings from voting multiple times. We can probably excuse individuals for participating in unreliable polls, since it's hard to resist an opportunity to register an opinion or vent frustration. The bigger question is why newscasters continue to conduct such polls and talk about their meaningless results, as they must surely know better. Perhaps in some cases it's just for fun or for ratings, but in the case of political commentators, I suspect it's more likely that they believe enough people are "bad at math" to be fooled.

Let's turn next to answer E, the special election. Democracy is a very good thing, and there's no better way to gauge public opinion than an election that generates large turnout. Unfortunately, it's tough to get high turnout even for presidential elections, and special elections tend to draw particularly small turnouts. Like online polls, these elections therefore tend to be biased toward those people with strong feelings on an issue (though we can hope they are only allowed to vote once). Moreover, if we are only trying to gauge public opinion (rather than make a binding decision), middle-of-the-road people are even less likely to vote than in other special elections. Combining those facts with the huge cost of holding a special election, we conclude that E would be a very poor choice.

The three remaining choices (A, B, and C) all involve random selection of much smaller numbers of people. You might at first think that choice C would be best, since it has the largest number of people. However, it should raise at least one red flag and suffers one more obvious problem. The red flag is that it is conducted by volunteers; unless they have been well trained, the volunteers could inadvertently (or deliberately) inject their own biases into the interviews. The more obvious problem is that even if the volunteers do a good job, Monday-morning interviews at grocery stores will tend to over-represent stay-home parents and underrepresent people who work a standard business week with an employer that provides health benefits. We therefore can't count on this poll to give results that are representative of the full spectrum of Americans. Choice B also suffers from bias, both because people who have recently been hospitalized tend on average to be older and less healthy than the rest of the population, and also because their experience may have affected their views. We conclude that choice A, the poll of only 1,000 people, is the one most likely to give an accurate assessment of the overall opinions of all Americans.

Even if you've known all along that A is the correct answer, it's still quite astonishing when you think about it. A survey of 1,000 people means asking only about 1 out of every 300,000 Americans; to use a perspective technique from the last chapter, this is like choosing just a single individual from six stadiums full of people. Viewed in this way, it seems almost impossible to imagine that the survey could be reflective of overall opinions. But it can be, if it is conducted carefully to ensure that the selection is truly random and not biased toward any particular segment of the population. Moreover, it's possible to quantify the uncertainties in a well-conducted poll. As you'll find posted on the Web sites of organizations like Gallup and Pew, a poll of about 1,000 people has a *margin of error* of less than about 4 percentage points,[6] which we can interpret as follows: If such a poll were repeated many times, each time with a different randomly selected group of 1,000 people, 95% of the polls would get within 4 percentage points of representing the true feelings of all Americans. A specific example should help. Suppose the poll finds that 580 of the 1,000 people, or 58%, say they support the health care plan. Then there is a 95% chance that this poll is within 4 percentage points—which in this case represents a range from 58% − 4% = 54% to 58% + 4% = 62%—of correctly

---

6. The formula for the approximate margin of error is $\frac{1}{\sqrt{n}}$, where n is the number of people polled, so for 1,000 people it comes out to be about 0.032, or 3.2 percentage points.

representing the views of all Americans. Based on this fact, we can say with "95% confidence" that the percentage of all Americans who support the plan is between 54% and 62%.

A famous story illustrates the remarkable power of careful polling. In 1936, editors of a magazine called the *Literary Digest* conducted a huge mail survey to predict the outcome of the upcoming presidential election. They mailed postcard "ballots" to some 10 million people that they chose at random from telephone directories and other lists available at the time. About 2.4 million people returned the ballots, and the results predicted that Republican candidate Alf Landon would win a landslide victory over President Franklin Roosevelt. In reality, Roosevelt won a landslide victory over Landon. In retrospect, the major problems in the *Literary Digest* poll were that the lists used for mailing postcards were biased toward more affluent voters (the only ones who could afford telephones back then), who tended to vote Republican, and that people who wanted a change were more likely to take the time to return the postcards than those who were satisfied with Roosevelt. The most interesting part is that a young George Gallup also conducted polls the same year. His surveys involved interviews with only about 3,000 people, barely one-thousandth of the number who responded to the *Literary Digest* poll. But thanks to his understanding of the principles of statistics, he correctly predicted the outcome of the election; he even conducted surveys that helped explain the incorrect outcome of the *Literary Digest* poll.

Opinion polls are only the beginning of the statistics that you see every day in the news. The latest data on the economy, new recommendations about when to get mammograms, reports on methods for improving education—they're all coming from statistical studies. We now live in the world that H. G. Wells predicted in his quote at the beginning of this chapter, in which statistical thinking has become as important to our roles as citizens as the ability to read and write.

In the rest of this chapter, I'll give you a brief overview of the major ideas you need to understand the statistics you encounter in daily life. First, however, we should clear up a common point of confusion about the word *statistics* itself, confusion caused by the fact that the word has a dual meaning. When we use it as a plural word, statistics *are* data that describe or summarize something; examples of data statistics include the percentages of people who say they support or oppose the health care plan in an opinion poll, batting averages and win-loss records in baseball, and economic statistics such as the unemployment rate and gross domestic product. When we use it as a singular word, statistics *is* the science of collecting, organizing, and interpreting all those data. To summarize, we use the science of statistics (singular) to help design statistical studies, and in the course of those studies we collect numerous pieces of data that are also called statistics (plural).

Truth, Truthiness, and Statistics

You've probably heard the famous line, "There are three kinds of lies: lies, damned lies, and statistics," commonly but incorrectly attributed to nineteenth-century British prime minister Benjamin Disraeli.[7] It's certainly true that statistics are often used in misleading ways, such as when people pick and choose only selected statistics to support a predetermined viewpoint. But this damning line about statistics applies primarily to the plural, data kind of statistics, which are relatively easy to take out of context. In thinking about the science of statistics, I prefer the phrase "truth, truthiness, and statistics," with thanks to Stephen Colbert for the second word. Let me explain.

A few philosophers notwithstanding, I think we can agree that *truth* is something that really exists. Whatever we may want to know, from the opinions of Americans on health care to whether the flu vaccine is safe and effective, there is a truth that is out there and waiting to be discovered. The question is how to go about discovering it. One far-too-common approach is through *truthiness*, which Colbert defines as "truth that comes from the gut." In other words, truthiness is the practice of deciding what's true based on intuitions or personal beliefs rather than actual facts. The problem with truthiness is that it may not accurately reflect the real truth that is out there; as another famous quote with obscure origins says, "It ain't so much the things we don't know that get us into trouble, it's the things we know that just ain't so."

---

7. Numerous earlier references show that Disraeli certainly was not the first to say it, and there's no definitive evidence that he ever said it at all.

This is where the science of statistics comes in. In most cases, there's no way to determine the truth beyond all doubt, at least within the limits of what we can realistically accomplish. For example, you can't actually check the opinions of all Americans, and you can't check for every possible effect of the flu vaccine on every human being. But thanks to the science of statistics, we have techniques that allow us to study relatively small samples of the full population and still learn what we'd *probably* find if we could study everyone. We can ask just one thousand people their opinions and be reasonably confident that the results reflect the opinions of all Americans, and we can test a new flu vaccine on just a few thousand people and make a reasonable judgment about its safety and effectiveness. Equally important, just as we saw earlier for the health care opinion poll, any well-conducted statistical study allows us to quantify our level of confidence (such as saying that we can be 95% confident) in its results.

I think you can now see what I mean by "truth, truthiness, and statistics": The truth is out there, and while people sometimes try to find it though the gut feelings of truthiness, it is the science of statistics that often provides our best hope of discovering the real truth.

How Statistics Works

Statistics today is a wide-ranging subject with applications that touch almost every aspect of our lives. For example, search engines like Google and Bing use advanced statistical formulas to organize search results, and intelligence agencies are constantly developing new data-mining techniques to try to find terrorists and guard against cyberattacks. However, most of the statistics that we encounter in the news come from studies that share major elements in common.

The Statistical Process

Figure 2 shows the basic process of a statistical study. The first step always is to identify the goals, which then help determine the relevant *population*. For

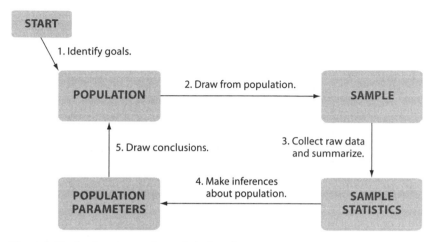

Figure 2. The basic process of a statistical study.

example, if your goal is to know where your candidate stands in an upcoming election, then the population should consist of all people likely to vote in the election; if your goal is to determine the effectiveness of a new cancer drug, the relevant population is all people who suffer from that type of cancer.

Once the population is identified, the second step calls for drawing a *sample* from the population. In the case of an opinion poll, the sample consists of the people who actually answer the question; in the case of a cancer drug study, the sample is the group of cancer patients who are monitored in the study. Sample selection is arguably the most critical step in any statistical study. If the sample is representative of the population, then there's a good chance you'll be able to draw valid conclusions even from a relatively small sample. But if the sample is somehow biased, as was the case with the *Literary Digest* survey, then your results will probably be invalid even if your sample is quite large.

The third step is to collect all the data from the sample and summarize the data into a set of *sample statistics*, by which we mean statistics (in the plural sense of summary numbers) that represent the actual findings for the sample. For example, in a poll asking 1,000 people whether they support the health care plan, the data consist of all 1,000 individual responses, but you need only one sample statistic—the percentage of the 1,000 people who support the plan—to summarize the results. The cancer drug study may require several sample statistics to summarize it, such as the average change in tumor size with the drug and without the drug, and the average number of months or years that the drug prolonged life.

The same statistics measured for the sample have counterparts in the population, usually called the *population parameters*. In the case of the health care opinion poll, for example, the population parameter would be the actual percentage of the entire population that supports the new plan, while in the cancer drug study the population parameters would be the results you would find *if* you could give the drug to everyone with the disease. As we've already discussed, we can never be certain that the sample statistics accurately reflect the population parameters, but if the sample was chosen well, then there's a good chance that they do. Therefore, the fourth step in the study process is to analyze this chance in detail, stating precisely what we can infer about the population parameters from the sample statistics, along with our level of confidence in those inferences. Although this step can be mathematically complex to carry out during the actual research process, it should be easy to interpret in news reports, as long as you are told the confidence level and any other uncertainties. For the all-too-common cases in which you are *not* told those uncertainties, you should recognize that critical information is missing before you take the results too seriously.

The final step in the process is to take everything you've learned and use it to draw conclusions related to the original goal and population. In the case of the opinion poll, the conclusion might be that the public supports or opposes the new plan, and in the case of the new cancer drug, the conclusion might be that its benefits are too small to justify its cost, or that it works so well that it should be made immediately available.

You might think that it would be straightforward to draw conclusions once you have the results, but the last step is probably the one most responsible for the association of statistics with lies and damned lies. After all, if someone wants to support a predetermined view, it's easy to "spin" the results of a report by selectively deciding which ones to emphasize. But the fact remains that if a statistical study is done well and the results are interpreted carefully and without bias, there is no better way to learn the truth about what is really going on.

Types of Statistical Study

Nearly all statistical studies follow the basic process I've outlined, but the details can vary significantly. Most statistical studies fall into one of two categories based on how they deal with the people (or animals or objects or whatever) that make up the sample drawn from the full population, which for convenience we'll call the *subjects* of the study. The first category includes opinion polls and other studies designed to learn something about the subjects of the study without trying to alter their behavior. These are usually called *observational studies*, because aside from things like asking questions, they do nothing more than *observe* what subjects say or how something affects them. The second category is *experiments*, which means studies in which a *treatment* is applied to some or all of the subjects, generally with the goal of learning the effects of the treatment.

Most studies are observational, both because they are generally cheaper and easier to conduct than experiments, and because they avoid potential ethical issues. For example, suppose you want to learn about the danger of talking on a cell phone while driving. Conducting an experiment would require asking at least some people who otherwise would not do so to talk on cell phones while driving, which would be unethical if you thought this could cause them to have accidents. Fortunately, you can also study this issue observationally; because many people already talk on cell phones while driving, you can simply compare their accident rates to those of people who don't talk on cell phones while driving.

The advantage of experiments, when they are possible, is that they can give more definitive results. To see why, imagine that there were some genetic trait that predisposed people both to being more sociable and to taking more risks. If such a trait actually existed (there's no evidence that it does), then we might expect such people to talk more on cell phones (because they are more sociable) *and* to be more risky drivers—in which case the cell phone users would have higher accident rates, but their genetic predisposition to risky driving rather than the cell phones would be the cause. Now, imagine instead that we conducted an experiment in which some of the subjects were randomly assigned to use a cell phone while driving, while others were told not to use a cell phone. Because we chose the groups randomly, any genetic predisposition to sociability and risk would presumably be present at roughly equal levels in both groups (as long as the group sizes are large enough). Therefore, a finding that those using cell phones had higher accident rates could not easily be due to anything besides the cell phones.

As the above example illustrates, many experiments can provide meaningful results only if we divide the subjects of the study into at least two groups: one group that receives the treatment (such as the "treatment" of using a cell phone) and a second group that does not; the latter group is usually called the *control group*. To get meaningful results, the members of both groups must be chosen at random, and both groups must be large enough to make it unlikely that they will have any substantial underlying differences.

Another important consideration in experiments will be obvious to anyone with kids, since you'll know that little owies are easy to cure with a kiss or anything else that makes a child believe that you've actually done something to help. In technical terms, the kiss or other "fake treatment" (a treatment that we don't expect to have a physical effect) is called a *placebo*, and the fact that placebos often produce real results is called the *placebo effect*. Although it's not too hard to see why the placebo effect would help with minor owies, careful studies have shown that it can be surprisingly powerful. One of my favorite stories of this power comes from a study conducted in the 1990s in which researchers were testing whether a drug (called Propecia) could stop or reverse balding. The subjects were all drawn from the population of men with male pattern baldness, and they were divided into a treatment group that received the real drug and a control group that received placebos, which in this case were just fake pills. Remarkably, the placebo at least temporarily stopped the balding of 42% of the men in the control group; in some cases, these men actually grew new hair, even though they weren't taking anything more than a sugar pill.

The power of the placebo effect means that it's very important to make sure that the people in an experiment don't know whether they are receiving the real treatment or a placebo; after all, fake treatments are more likely to work if you believe they are real. In statistical terminology, we say that the study should be conducted *blind*, meaning that the subjects are blind as to whether they are in the treatment or the control group. If the treatment group has a significantly higher response rate than the control group,[8] then you can be confident that the treatment has a real effect that goes beyond the placebo effect.

In some cases, the experiment needs to go further and be *double blind*, meaning that neither the subjects nor the people administering or evaluating the treatments know which people are receiving the placebo. As an example, consider a study of the effectiveness of a drug for children with attention deficit disorder (more technically called attention deficit hyperactivity disorder, or ADHD). Doctors rely on observations of a child's behavior in order to diagnose this disorder, and observers who believe that the drug is effective might have an unintentional tendency to record behavior differently for children receiving the drug than for those receiving the placebo. The only way to prevent such tendencies from influencing the results is to make sure that the observers don't know which children receive the real drug and which receive the placebo.

The key point to remember is that the *details matter* in how a study is conducted. Observational studies are conducted differently from experiments, and experiments must be done with special care to avoid problems like those caused by the placebo effect. Therefore, one of the first things you should do upon hearing results from any statistical study is to make sure you know how the study was conducted, so that you can decide whether you think its results are likely to be believable.

Statistical Data in the News

If you were enrolled in a statistics class, at this point we would turn our attention to more detailed study of methods for calculating the summary statistics and confidence levels. Our goal here is more modest: I'd like you to feel

---

8. A key issue is defining what constitutes a "significantly higher" response rate; generally speaking, we say that a statistical result is *significant* if it has less than a 5% probability of being due to chance (rather than to the treatment under study). Statisticians say this is significance at the "95% level" (or sometimes at the "0.05 level"). Of course, for many studies we might demand much higher levels of significance before believing the results.

you can make intelligent decisions about whether to believe the results of the many statistical studies you hear about in the news. The next step in helping you achieve that goal is to examine a few of the most common issues that arise in news reports of statistical data.

Average Confusion

We all know that people's incomes suffered during the recent recession, but how were incomes doing before that? You might think that this question would be easy to answer just by seeing how average incomes have changed over time. However, if you dig through the statements made by political pundits, you might find both of the following claims:

- Adjusted for inflation, income has been relatively stagnant over the past three decades, with average household income in the United States rising less than 10% since 1978.
- Adjusted for inflation, income has risen significantly over the past three decades, with average household income in the United States rising more than 27% since 1978.

Wait, you say: the two statements are contradictory, so at least one must be wrong. But in fact, both are true; they simply use different definitions of the term *average*. The first statement is based on what we call the *median*, while the second is based on the *mean*. Let's see how they give such different results.

Suppose there were exactly 100 million households in the United States. (The actual number is a bit higher.) Imagine making a list of all 100 million household incomes, ranked from lowest to highest. The *median* is simply the middle value in the list, so 50 million households have incomes below the median and the other 50 million have incomes above it. Finding the *mean* requires a calculation: You would add up all 100 million incomes to find the total income of all the households, then divide by 100 million to find the *mean* income per household.

Perhaps because it requires computation, schools tend to spend more time teaching kids about the mean than about the median; as a result, most people think of the mean when they hear the word "average." But the mean and the median are both legitimate definitions of "average," and there are many cases in which the median is the better choice. As an extreme example, imagine that ten seniors on a college basketball team all hope to get NBA offers, but only one does. This one player is given a contract offer for $10 million, while the other nine players get nothing. If we make a list of the 10 contract offers in order of amount, it looks like this: $0, $0, $0, $0, $0, $0, $0, $0, $0, $10 million. Therefore, the *median* contract offer is zero, since that is the middle value in the list. In contrast, the *mean* contract offer for the 10 players is $1 million, which we find by taking the total amount of their contract offers ($10 million) and dividing by 10. The basketball coach would probably like to tell potential recruits that the "average" senior gets a $1 million offer from the NBA, but the median of zero is a more realistic measure of what most players can expect.

As the basketball example shows, a small number of high values can make the mean much higher than the median. More generally, the mean will be higher than the median for any distribution that is skewed (lopsided) toward high values. Household incomes are skewed in this way, because the top few percent of households earn a very large proportion of total income. For example, the mean household income in the United States (as of 2010) is about $68,000, while the median is only about $50,000.

We can now apply these ideas to the two bulleted statements about income. The first statement, based on median income, tells us what happened to income in the middle of the income distribution; the fact that it rose only a little less than 10% tells us that most households saw relatively little income gain since 1978. Because the second statement, based on mean, shows a much larger gain of 27%, we conclude that income gains were much higher for households at the higher end of the income scale.

Not surprisingly, people tend to use the definition of "average" that works best with their personal beliefs. With incomes, conservatives tend to focus on the mean because it is higher and arguably a better measure of the overall economy, while liberals tend to focus on the median because it is lower and arguably a better measure of what is happening to most people. Similarly, in a labor dispute, workers will tend to cite their median wages when arguing for a pay raise, while employers will prefer the mean since it is usually higher.

In fact, mean and median aren't the only legitimate definitions of the term "average," though others are less common. For example, the most common value in a distribution, technically called the *mode*, is occasionally cited as the average. And if you recall grading schemes in school, you'll be familiar with what is sometimes called a *weighted average*, in which the final exam counts for more than the midterms when your exam grades are "averaged" to find your final grade.

Given the various possible definitions, you can be easily misled if you are thinking of one kind of average while a statement is referring to a different one. In most cases, however, a little digging will tell you whether the stated average is a mean, a median, or something else. Once you know that, you can then decide for yourself whether you think the average being used is appropriate for the situation.

Pictures of Statistics

When we see statistical data in the news, they are almost always in the form of either tables or pictures. The pictures can be of many types, such as pie charts, bar graphs, and line graphs. These pictures are usually designed to be fairly easy to interpret, but on occasion they are made in ways that can be deceptive if you don't think carefully about them. Let's look at a few of the techniques that often lead to misinterpretation.

For our first example, consider Figure 3, in which dollar bills get smaller to indicate their declining value over time (due to inflation). Although figures using this visual technique are quite common, they tend to make the changes seem greater than they really are. The problem is that the values are represented by the *lengths* of the dollar bills; in this case, a 2010 dollar was worth $0.39 in 1980 dollars and therefore is drawn so that it is 39% as long as the 1980 dollar. However, our eyes tend to focus on the *areas* of the dollar bills, and the area of the 2010 dollar is only about 15% of the area of the 1980 dollar.[9] In other words, unless you are careful to focus on the lengths, you're likely to end up thinking that the change was much greater than it was.

---

9. Areas always scale as the square of length; for example, an 18-inch pizza has $2^2 = 4$ times as much total pizza as a 9-inch pizza. That is why the area of the 2010 dollar is $0.39^2 = 0.1521$ times (or 15.21%) that of the 1980 dollar.

1980 = $1.00   1990 = $0.63   2010 = $0.39

Figure 3. The dollar *length* represents its declining value with time, but our eyes tend to focus on the even greater change in area.

Women as a Percentage of All College Students

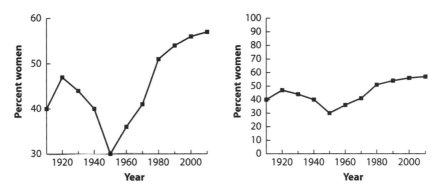

Figure 4. Both graphs show the same data, but they look quite different because they use different scales on the vertical axis. Data from the National Center for Education Statistics.

A similar exaggeration can occur when a graph is shown with an expanded scale. Figure 4 shows two graphs of the percentage of women among college students. Both graphs show the same data, but the one on the left makes the change look much greater, because its vertical scale goes only from 30% to 60% rather than from 0% to 100%. There's nothing inherently wrong with this technique; in some cases, expanding the scale is the only way to make a change visible. But you can also see how important it is to pay close attention to the scales, since otherwise you'd think the two graphs were showing very different data.

Next, look at Figure 5, which shows how computer speeds have increased with time. If you didn't look carefully, the straight line in the graph on the left might make you think that computer speed rose by the same amount in each decade. But this is not the case, because each tick mark along the vertical axis

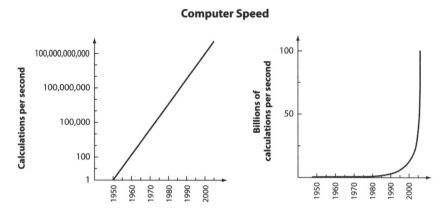

Figure 5. Both graphs show the same data, but you can read the data easily only from the left graph, which uses a vertical scale in which each tick mark represents 10 times as many calculations per second as the prior one.

represents a speed 10 times as great as the one below it. When we remake the graph with a normal (linear) scale, we get the graph on the right, in which you can see that the increase in computer speed becomes much greater with time. The advantage of the left scale (sometimes called an *exponential* or *power* or *log* scale, because the tick marks rise by powers of 10) is that, at least in this case, it makes it much easier to read details. For example, the left graph shows clearly that the speed rose from about 100 calculations per second in 1960 to about 100 billion calculations per second in 2000, while the right graph makes it hard to see anything other than that it changed a lot. Both scales are legitimate and have their uses; you just need to make sure you understand which type of scale you are looking at.

For our final example, Figure 6 shows graphs of changes in college costs. If you didn't look too carefully, the top graph might lead you to conclude that after peaking in the early 2000s, the cost of public colleges fell during the rest of the decade. But the vertical axis is showing the *percentage change* in cost each year, so the drop-off means only that costs rose by smaller amounts, not that they fell. Actual college costs are shown in the graph on the bottom, which makes it clear that they rose every year. Graphs that show percentage change are very common; you'll find them in the financial news almost every day. But as you can see, they can be very misleading if you don't realize that they are showing change rather than actual values.

Changes in College Costs

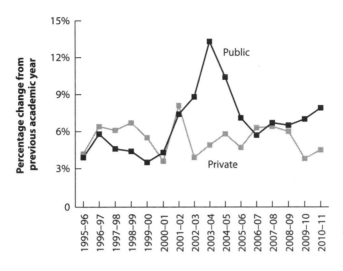

Actual College Costs

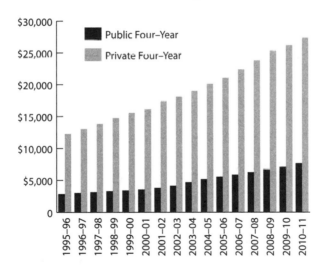

Figure 6. Did college costs fall in the late 2000s? The top graph might make it look that way, but it's actually showing only decreases in the rate at which costs rose. The bottom graph shows the trend in the actual costs. Data from the College Board.

Correlations and Causality

Many statistical studies are designed to learn whether one factor *causes* another. We may want to know if cell phone use causes more car accidents, if a new drug causes patients to get better, or if microlending programs cause improvements in the lives of people in developing nations. But we need to use caution in looking for causality, and in particular, we need to distinguish it from simple correlation.

A bit of terminology will help our discussion: Anything that can vary or take on different values is called a *variable*. In algebra, we typically use the letters *x* and *y* as variables, meaning that they can represent many different numbers. For a statistical study, we usually identify *variables of interest*, meaning the things that we actually intend to measure. In a study of childhood obesity, the variables of interest would include age, gender, height, and weight for all the children in the study. In a study of cell phones and driving, the variables of interest would include accident rates with and without cell phone use.

We say that there is a *correlation* between two variables whenever a change in one variable is accompanied by some consistent change in the other. The variables height and weight are correlated in people because taller people tend to weigh more, at least on average. Height and weight have a *positive* correlation, because both variables tend to rise or fall together. Correlations can also be *negative*, meaning that one variable rises while the other falls; price-demand relationships provide an example, since an increase in price usually leads to a decrease in demand.

Statistical studies are very good at finding correlations, but a correlation by itself is never enough to establish causality. For example, numerous studies have established a clear correlation between cell phone use and car accidents, but as we discussed earlier, that does not necessarily mean that the cell phone use is the cause of the accidents (since we can imagine other possibilities, such as a gene that predisposes people to both sociability and risk taking). In general, a correlation between two variables can occur for any of three reasons: (1) The correlation may be a coincidence, meaning there really is no relationship between the two variables at all; (2) the correlation may exist because the same underlying cause affects both variables; or (3) one variable may actually be all or part of the cause of the other.

A classic example of a coincidental relationship is the "Super Bowl Indicator," a correlation between the winning team in the Super Bowl in January and stock market performance for the rest of the year. The correlation is this: When the winning Super Bowl team comes from the old, pre-1970 NFL, the stock market tends to rise during the rest of the year; otherwise, it tends to fall. Remarkably, the Super Bowl Indicator successfully "predicted" the stock market direction based on the winners of twenty-eight of the first thirty-two Super Bowls, a track record far better than that of any professional stockbroker. In fact, detailed calculations show that the probability of such a good track record arising by chance is less than 1 in 100,000, so small that by standard statistical measures we'd presume there's something real going on here. But chance it still is, and no one in his or her right mind would use the NFL origins of the Super Bowl winner to determine an investment strategy for the next year. More recent events bear this out; after Super Bowl 32, the Super Bowl Indicator successfully predicted the stock market direction in only five of the next ten years, which is exactly the fraction that would be expected by pure chance.

For an example of the second explanation for a correlation, consider the fact that cities with more churches also tend to have more liquor stores; that is, there is a strong correlation between the number of churches and the number of liquor stores in cities. While an atheist might be tempted to conclude that churches drive people to drinking, the more obvious explanation is that the two numbers rise in tandem because they share a common underlying cause: population. A larger population leads to both more churches and more liquor stores.

The third explanation, causality, is the most interesting, because it can lead us to make decisions that can have a real impact. Smoking rates have declined, and smoking has been banned in many public places, because statistical studies showed so clearly that both smoking and secondhand smoke can cause lung cancer and other diseases. But given the other two possible explanations for a correlation, how do we go about establishing that one thing actually causes another?

Researchers use a number of general guidelines in establishing causality, but for our purposes it will be clearer to consider a couple of examples. Let's start with the case of air bags and children. When automobile manufacturers began adding air bags to cars, everyone assumed that they would save lives. Accident and fatality statistics quickly confirmed that air bags did indeed save lives in moderate- to high-speed collisions, but an unexpected and disturbing pattern also appeared: In low-speed collisions, the air bags were correlated with an *increase* in deaths among young children.

At first, the correlation seemed to make no sense, and many people assumed it was due to coincidence. But the correlation persisted in further studies, making it less and less likely that coincidence could be responsible. Researchers therefore examined the data more closely and discovered that the correlation arose only for children in front seats, not back seats, and that the risk was greater for smaller children and became even greater when those children were in booster seats or infant car seats. Still somewhat mystified, researchers conducted crash tests with child- and adult-size dummies. These tests showed that the explosive openings of air bags tended to be directed at heights that were safe for adults, but that aimed squarely at the vulnerable heads and hearts of smaller children. Children in boosters and infants in car seats were at even greater risk because their seats put them closer to the air bags. Once researchers developed this physical understanding of the reason for the correlation, there was little room left for doubt in saying that air bags could cause death among young children in front seats. That is why safety advocates now recommend that children under twelve (and older children shorter than 4'9") sit in back seats at all times.

The issue of cell phones and driving has followed a similar trajectory. The correlation between cell phone use and accidents was also thought to be coincidental at first, since talking on a cell phone doesn't seem to be so different from talking to a passenger, and the latter did not appear to be correlated with accident rates. As evidence for the correlation accumulated, the idea of coincidence became harder and harder to accept, so researchers began to look for other explanations, and one seemed obvious on the surface: most cell phone users were holding the cell phones in their hands, which seemed likely to mean less control while driving. This led to calls for laws requiring hands-free devices in cars. However, while those devices probably can't hurt, further studies showed that they didn't make the correlation go away; that is, it was the act of talking on the cell phone that was correlated with accident rates, not the act of holding the phone.

At this point, researchers began to question the assumption that talking on a cell phone is just like talking to a passenger. Brain scans conducted during simulated driving sessions soon showed that talking on a cell phone activates different areas of the brain than talking with someone sitting next to you. This offered a potential explanation for why talking on a cell phone might have different effects than talking to a passenger. Follow-up studies provided additional evidence for the idea that cell phone use is a source of distraction that can cause accidents with or without hands-free devices. Most shockingly, some studies have found that the distraction caused by talking on a cell phone while driving—even with a hands-free device—can make you as dangerous as a drunk driver. Texting or using other computing devices while driving is even worse, because it causes the same type of distraction but also forces your eyes off the road. The National Safety Council now estimates that approximately 1.6 million car crashes each year, more than a quarter of the total, are caused by some type of distraction.

To sum up this discussion, you should focus on two main points. First, it's not easy to establish causality. Finding a correlation is only the first step in a long process, and we can be confident in causality only after careful and in-depth study. Second, where we *can* establish causality, knowing the causal relationship gives us a strong basis for making decisions. For example, although we may not yet have political agreement on what to do about cell phones and driving, your personal decision should be easy: Hang up the phone. Knowing that cell phone use, texting, and other distractions make you as dangerous as a drunk driver means there's just no justification for them. If there's a call you just can't miss or a message you just must send, or you just have to put a new destination into your GPS, then find a place to pull off the road.

The Bell Curve

For our last general topic in statistics, let's turn to the famous *bell curve*, best known through teachers who "grade on the curve" and for its role in the IQ debate. In statistics, the bell curve is more formally known as the *normal distribution*, shown in Figure 7; the nickname comes from the characteristic bell shape. The peak of a normal distribution is always located at its mean (average)

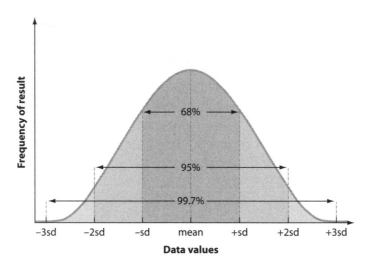

Figure 7. The normal distribution has a bell shape. Data values are clustered near the mean, and the width of the bell—which tells us how far data values tend to be spread around the mean—is described by the standard deviation (sd).

value.[10] Other data values are spread symmetrically around this mean, with values farther from the mean being less common.

The normal distribution is important because it arises often. For example, many human characteristics show a nearly normal distribution, including height, weight, hundred-meter dash times, and calories consumed per day. The normal distribution is also common for test scores (as long as the mean score is not too close to 0% or 100%), which explains its popularity in grading.

The two key characteristics of any normal distribution are the location of its peak, which is always at the mean, and the amount of spread in the distribution, which we describe with a number called the *standard deviation*. A large standard deviation implies that data values are spread out over a wide range, giving the distribution the shape of a wide bell. A small standard devia-

---

10. To connect with our earlier discussion, notice that the median and the mean are equal for a normal distribution because of its symmetric shape. The distribution of incomes is *not* normal, because people with very high incomes cause the shape to be skewed to the right, which is why the mean is higher than the median in that case.

tion implies that data values are concentrated near the mean, giving the distribution the shape of a narrow bell. In all cases of normal distribution, data values are distributed around the mean according to the rules shown in Figure 7: About 68% (just over two-thirds) of all data values lie within one standard deviation of the mean, about 95% lie within two standard deviations, and about 99.7% lie within three standard deviations.

To make the ideas clearer, let's take the specific example of IQ scores. If you make a graph of IQ test results for thousands or millions of people, you'll see a normal distribution. To make these results easy to interpret, each actual (raw) test score is converted to an IQ, which is defined so that the mean IQ is 100 and the standard deviation is 15. Therefore, about two-thirds of the population has an IQ within 15 points of the mean of 100, or between 85 and 115; about 95% of the population has an IQ within $2 \times 15 = 30$ points of the mean, or between 70 and 130; and about 99.7% of the population has an IQ within $3 \times 15 = 45$ points of the mean, or between 55 and 145.

You can interpret other normal distributions similarly. For example, the heights of American women are normally distributed with a mean of about 65 inches and a standard deviation of about 2.5 inches. Therefore, about two-thirds of American women are between about 62.5 and 67.5 inches tall, 95% are between 60 and 70 inches tall, and 99.7% are between 57.5 and 72.5 inches tall. If you are a woman who is shorter than 57.5 inches or taller than 72.5 inches, then you are in groups that, combined, represent only about 0.3% of the population; that is, about 0.15% of women are shorter than 57.5 inches and another 0.15% are taller than 72.5 inches.

Normal distributions lurk behind the scenes of most of the statistical studies you hear about in the news. Consider the widely reported study (released by the Kaiser Family Foundation in 2010) that found that the average child spends about seven and a half hours a day using electronic devices such as computers, iPods, and smart phones. It's safe to presume that this average represents the mean of a normal (or nearly normal) distribution, a fact that should give you a better mental picture of what the data really look like. Indeed, you'd have an even better picture if you were told the standard deviation; unfortunately, news releases to the public rarely provide this number, perhaps because media officials think the term "standard deviation" is just too scary for the average person. I hope someday we'll prove them wrong.

Should You Believe a Statistical Study?

We're now ready to take the basic ideas of statistics that we've discussed and put them together to meet our key goal: being able to decide whether to believe the results of a statistical study reported in the news. There are many ways to go about this decision process, but I'll present you with eight simple guidelines that should help.

Get a big-picture view of the study. The first thing you should do when you hear about a statistical study is figure out what it was all about; in particular, you should understand the goal of the study, the population that was under study, and whether the study was observational or experimental. This big-picture view will help you put the results of the study into context, and it may sometimes turn up obvious limitations. For example, a medical study on a population of white males cannot be assumed to apply equally to women or to men of other ethnic groups, and a pre-election opinion poll that doesn't distinguish between the populations of registered voters and likely voters will likely give poor results.

It's especially important to decide whether the type of study was appropriate. Consider a study in which people are given personalized horoscopes and asked if they are accurate (which makes it an observational study, since the researchers observe but don't try to alter the responses). Such studies usually find most people saying their horoscopes are accurate. But these results are meaningless for several reasons, including the fact that most horoscopes are so general that their accuracy is highly subjective. The best way to get a meaningful result about the validity of horoscopes is through an experiment, such as giving some people their actual horoscopes while giving people in a control group horoscopes that belong to someone else (so that these "other" horoscopes serve as the placebo). Such experiments have been done many times, and they have never turned up a significant difference in how the two groups rate the accuracy of the horoscopes. In other words, the experiments allow us to conclude that horoscopes lack validity, regardless of what any observational study might say.

Although the horoscope example may seem rather obvious, similar problems crop up all the time. According to the National Institutes of Health, Americans now spend more than $30 billion per year on "alternative medicine" treatments such as homeopathy, chiropractic, and crystals, in part because the marketing of these treatments usually emphasizes observational studies in which people have reported benefits. But, as with horoscopes, the best way to establish whether these treatments work is through experiments, and experimental evidence is distinctly lacking. Of course, absence of evidence is not evidence of absence, and it's possible that some of these treatments may actually work. I've even tried some of them myself, and I have to admit that in some cases they've seemed to help. Nevertheless, the statistical evidence to date is consistent with nothing more than a placebo effect, so at the risk of offending friends and readers who practice alternative medicine, I'll tell you what I really think: Unless and until much stronger evidence comes along, I think we could find better uses for that $30 billion per year.

Note that mainstream medicine can suffer similar problems. Sometimes experiments may be impractical, even though they are the only way to get clear results. For example, surgical techniques are difficult to test through experiments, because the placebo has to be a fake surgery that makes people in the control group believe they've had a real surgery. Although such experiments have occasionally been conducted, the inherent danger of any surgery (real or fake) means these studies are fraught with ethical concerns; as a result, the evidence for many common surgical techniques is not nearly as strong as you might guess. Even in cases where an experiment is possible, there may be problems with the way control groups are defined. Consider an experiment that demonstrates that a new drug is more effective than a placebo. Although such a study seems to establish that the new drug works, the more relevant issue may be whether the new drug is better than an older (often less expensive) drug. Unless that older drug was also tested in the experiment, you can't draw a conclusion on that issue.

Consider the source. True story: On the day I started writing this section, the headlines reported a new study showing that even a relatively small reduction in salt intake could have significant health benefits. The news reports also noted that not everyone accepted this conclusion; the Salt Institute, which is funded by salt producers, provided a spokesperson (no joke—his first name was Morton!) who expressed skepticism about the results and pointed to other studies that had not found such a clear link.

Situations like this arise in the news all the time, presenting us with conflicting claims and conflicting studies. Another recent debate has centered on whether women younger than fifty should get regular mammograms; while studies have shown that such mammograms clearly save some lives, other studies have suggested that the risks (from false positives and the radiation in the mammograms) outweigh the benefits for these younger women. Given that even the experts argue about which study to believe, how are the rest of us supposed to make a decision when all we have are news reports?

It's not easy. The salt study case may seem obvious, since we expect a salt industry group to be biased against anything that suggests reducing our use of salt. But just as being paranoid doesn't necessarily mean that people aren't out to get you, the fact that a group is biased doesn't necessarily make it wrong. Indeed, most statistical studies are funded by groups with bias, primarily because they're usually the ones most interested in the results. That is why most tests of new drugs are conducted by their manufacturers, and why milk producers sponsor studies in which they hope to establish health benefits for milk.

Still, you can use some common sense to help make reasonable judgments. A study conducted by the Salt Institute may be perfectly valid, but if it conflicts with a similar study conducted by independent researchers, then I'd be more inclined to trust the latter, as long as those researchers have no obvious bias. Similarly, if a pharmaceutical company funds a study of a new, expensive drug, I'm more inclined to believe the results if independent researchers, rather than the company's staff doctors, conduct the study. There's no foolproof way to know if a study has been done well and without the influence of bias, but considering the source is a good place to start.

Look for bias in the sample. We've already discussed some of the ways that bias can arise in sampling, but the issue is so important that it has to be included in this list of guidelines. Remember this key point: A statistical study can yield meaningful results *only* if the sample is chosen without bias and accurately represents relevant characteristics of the overall population. Although there are exceptions, getting an unbiased sample usually requires random selection.

Of course, even a well-chosen sample could by pure chance end up unrepresentative of the population. This is especially true if the sample is small. It's amazing how often you'll see news stories in which reporters interview, say, seven people, and then act as though these seven might somehow represent all Americans. As a rule of thumb, you should expect observational studies to include at least several hundred people, and experiments to include at least several dozen in each of the treatment and control groups. Larger sample sizes won't necessarily give better results, but they increase the probability that the results are accurate, as long as the samples are chosen without bias.

Look for problems defining or measuring the variables of interest. It's relatively easy to do statistical studies on IQ, but establishing that IQ is a valid measure of intelligence is another matter; indeed, there is great debate among psychologists about the extent to which IQ measures innate intelligence versus something learned. So if a study purports to tell you something about human intelligence, you should immediately wonder how the researchers defined and measured it. The same idea holds for studies on other variables of interest that cannot be easily defined or measured, such as love, happiness, and self-esteem.

Even if a variable is easy to define, it may be very difficult to measure. Years ago, I read an article in the *New York Times* that discussed a commonly quoted statistic: that law enforcement officers stop only 10% to 20% of the illegal drugs entering the United States. The article pointed out that while it's easy to measure the amount of illegal drugs collected by law enforcement, it's rather more difficult to measure the quantity of drugs that we *don't* intercept. An officer quoted in the article said that he and his colleagues refer to this type of statistic as "PFA," for "pulled from the air," a term I've appreciated ever since. You may not always have a way to know for sure whether a statistic is real or PFA, but at least be on the lookout for PFA warning signs.

Beware of confounding variables. For decades after scientists first began to link smoking with lung cancer and other diseases, researchers at the Tobacco Research Institute (consider the source!) disputed the claims. Given that lung cancer rates are clearly higher among smokers, you might wonder how they were able to dispute the link with a straight face. The answer is that they always pointed to what statisticians call *confounding variables*—variables that can affect results but that studies have failed to take into account. In the case of smoking and lung cancer, for example, confounding variables would include other possible causes of lung cancer, such as long-term exposure to radon or asbestos. Because early studies did not track exposure to these other variables, the Tobacco Institute researchers could claim with at least some legitimacy that the correlation did not establish causality.

To put this in the context of our earlier discussions, confounding variables are the leading cause of the difficulty of establishing causality. Recall that finding a correlation is only a first step in establishing causality. Because it's very difficult to be sure you've accounted for every possible variable that could affect a correlation, it's equally difficult to be sure that the correlation isn't spurious or due to some common underlying cause. That's why it's so important to study correlations very carefully before drawing conclusions about causality, and why it is so helpful if you can identify a physical mechanism that explains a causal link. Incidentally, in case you still harbor doubts, the physical mechanism by which smoking causes lung cancers is now well understood and has been observed in laboratory cell cultures. While it's never possible to prove causality beyond *all* doubt, the claim that smoking causes lung cancer has certainly been verified far beyond any reasonable doubt.

Consider the setting and wording in surveys. First question: Do you support or oppose President Obama's policies on dealing with radical Islam? Second question: Do you support or oppose President Barack Hussein Obama's policies on dealing with radical Islam? Third question: Do you believe that an opinion poll asking the second question would yield the same results as one asking the first? As you might guess, the mere act of inserting President Obama's middle name into the survey question can change the survey results, regardless of how well the sample is chosen.

More generally, research has shown that survey results can be very sensitive to wording. A question as simple as "Do you lean more toward the Republican Party or toward the Democratic Party?" can give different results if you change the order in which the two parties are named. For that reason, professional polling organizations usually randomly change the order of the party names in questions like this one, so that some people in the sample get one version and others get the second version.

Greater changes in wording can cause even greater differences, a fact often used by politicians and interest groups in "push polls" that pretend to be legitimate but really are pushing some viewpoint. Consider the issue of whether we should raise taxes. Republicans might ask, "Do you favor giving the government more of your hard-earned money through a tax increase?" while Democrats might ask, "Would you accept a tax increase to reduce the sky-high deficits that will crush our children?" Each poll will likely turn up the results that the party desires.

The setting of a poll can also affect results, particularly if the poll is on sensitive topics. If pollsters go door to door and ask people if they've ever cheated on their income taxes, they're likely to find more people saying no than if they ask the same question through some mechanism that ensures anonymity.

Check that results are presented fairly. In my home state of Colorado, officials have proudly pointed out that our public school students score as "proficient" on standardized state tests in reading and mathematics at a higher rate than those in almost any other state. Explanations commonly offered for this strong performance include the quality of our teachers, the strong curricula, and the relative affluence of our population. But there's a more likely explanation that gets relatively little press: Our state tests are easy. The U.S. Department of Education has compared state proficiency tests to the National Assessment of Educational Progress, which is sometimes called "the nation's report card." These comparisons show a wide variation in the difficulty of the tests given in different states, and it turns out that Colorado's tests are among the easiest. This fact is well known to education officials, but it doesn't seem to stop them from touting the state results anyway. You know we have a "bad at math" problem when even education officials can't present their statistical data fairly.

Of course, the problem of unfairly presented results goes far beyond education. Statistical studies often lead to conclusions that don't support people's self-interest or preconceived beliefs. In those cases, the people who report on the studies have a simple choice: They can report the results honestly, revealing an unexpected or uncomfortable truth, or they can spin the results into the realm of truthiness. You know the rest: As Mad-Eye Moody of the Harry Potter series reminds us, we need Constant Vigilance against those who conspire to corrupt truthful statistics with lies and damned lies.

Stand back and consider the conclusions. Our last guideline brings us full circle back to our first, the big-picture overview of the study. Imagine that, after thinking about all the other guidelines, you conclude that a study has been conducted well and its results reported fairly. The final step is to go back to your big-picture view of the study and decide whether the study's conclusions are really meaningful.

This last step can take many forms. Sometimes the issue may be one of practical significance. Consider the question of whether zinc lozenges can cure colds. Although most studies have found no effect from the zinc, at least one widely reported study found that, if taken starting with the first sign of a cold, the zinc reduced the average length of the cold from about seven days to about six and a half days. This explains why packages of zinc lozenges often say they are "proven to reduce the duration of colds." But knowing that too much zinc can be dangerous to your health, is it really worth even minimal risk for a half-day change in a cold's duration?

In other cases, it's a matter of asking whether the results make sense. Think back to the Super Bowl Indicator for the stock market. As we discussed, by standard statistical measures, the indicator's results for the first 32 Super Bowls would have led you to the conclusion that it's about the best predictor of stock market performance around. But it doesn't make sense, which is why no one ever took it seriously.

This last point is especially important regarding claims that push the bounds of established knowledge. For example, there are well-respected psychologists who have claimed that their studies show evidence that extra-sensory perception (ESP) really exists, and there are physicists who claim to have measured tiny deviations from the accepted laws governing gravity. It's always possible that these researchers really are on to something, but until they manage to convince a large number of their equally smart colleagues, I'd take it all with a grain of salt. As famed astronomer Carl Sagan used to say, "Extraordinary claims require extraordinary evidence."

Math for Life

Look back at the two quotations that open this chapter. By now, I hope I've convinced you that H. G. Wells was correct about the importance of statistical thinking. So it's the second quote I want to focus on now. As an educated reader, it probably made you laugh. But if you give this quote to elementary school children, you'll find that many of them don't get it right away, and for some you have to explain it clearly before they'll understand why it's funny.

I believe there's an important lesson here. Even things that may seem obvious are not always easy to understand, especially when you've never confronted them before. Most of what we've covered in this chapter probably seems fairly obvious in retrospect, but it's not at all obvious to most high school and college students. That's why one of my mantras in teacher workshops, whether for elementary teachers or for college faculty, is that *we can't expect students to know what they've never been taught*. Sure, there are some exceptional students who will figure these things out for themselves, but it's unreasonable to expect all students to do that. Given the importance of statistics in the modern world, there are real consequences of the fact that so few of our fellow citizens have ever had the opportunity to learn about statistics even at the level in this book, let alone at the more detailed level that they would get in a course in quantitative or statistical reasoning.

With that, I'll hop down from my soapbox and tell you where we go from here. In these past two chapters, I've given you an overview of what I consider to be the most important general principles of quantitative reasoning—those of thinking with numbers, and those of statistical thinking—in hopes that they will help you understand many of the topics you will encounter in your daily life. In the rest of this book, we'll switch gears from the general to the specific, with each chapter focused on a topic that I believe to be both important and well suited to the study of "math for life."

4

Managing Your Money

A fool and his money are soon parted.
— English proverb

If it sounds too good to be true, it probably is.
— Source unknown

Question: You're a high school graduate and figure you could use a million bucks. Your best strategy for getting it is:

Answer choices:
a. Wait for the lottery to have an unusually large prize, then buy a lot of tickets
b. Develop your athletic skills in hopes of becoming a professional athlete
c. Go to college
d. Invest in the stock market
e. Get a restaurant job, in hopes that you can move up through the ranks to management

After the general topics we've discussed in the past couple chapters, I imagine this one might grab your attention. After all, who wouldn't want an extra million dollars?

You probably recognize that choice A is there just to see if you've been paying attention; if not, look back to the stadium lottery example in Chapter 2, and you'll see why playing the lottery is never a smart financial strategy.

Choice B may be more enticing if you happen to be a gifted athlete, but a few statistics should dampen your enthusiasm. In men's basketball, for example, the National Collegiate Athletic Association (NCAA) reports that only about 3 in 10,000 seniors on high school basketball teams will end up being drafted by a professional team. The odds for football are only a little better, at about 8 in 10,000. The chances of becoming a professional are even smaller for most other sports, including track, swimming, tennis, and golf. Overall, there are only about 10,000 professional athletes in the United States, or less than 1 in 30,000 of us, which means your odds of becoming one aren't much different from the abysmal odds of winning the lottery. For 99.997% of us, sports are for fun and health, not for money.

We can also use statistics to evaluate choice C. The U.S. National Center for Education Statistics reports that the median income of people with only a high school diploma is about $28,000 (as of 2008), while the median for people with a bachelor's degree is about $53,000. That's a difference of about $25,000 per year. If we assume a 40-year career (say, from age 25 to age 65), that amounts to a total difference of 40 × $25,000 = $1,000,000. Bingo! Spending four years getting a college degree will "win" you $1 million over your lifetime, at least on average.[11] You'll generally earn more if you major in a high-demand field such as math, science, or engineering, and less in fields such as communications, humanities, and social sciences. Within any particular major, studying harder and getting better grades will also tend to increase your future earnings. Still, no matter how you look at it, going to college offers by far your best chance of eventually ending up with an extra million dollars, generally making it worthwhile even if you have to borrow to do it. A further benefit has become clear during the recent recession: Although unemployment is high for everyone, college graduates are only about half as likely to be out of work as those without college degrees.

Now that we've identified the correct answer, we know that D and E are incorrect. The recent gyrations of the stock market make it obvious that D

---

11. Of course, I have not included the cost of your college education or the fact that you may have given up earnings from a job during the time you attended college, both of which would tend to reduce the "extra" you earn by going to college. On the other hand, most college graduates begin work at ages younger than 25 (and white-collar workers tend to retire later than blue-collar ones), which would tend to increase the "extra."

carries a lot of risk, and it would take extraordinary success for you to parlay, say, the typical cost of college into a $1 million market gain. The problem with E is that, unless you are very lucky, the highest-paying management jobs will go to college graduates, which takes us back to answer C.

I chose this question to start our chapter on finance for one major reason: it shows the importance of *investing in yourself*. Many people seek a quick path to riches, but few ever find one. In contrast, those who take the time to invest in themselves, working hard to become more educated, are as close to guaranteed success as it is possible to get.

The importance of investing in yourself goes far beyond college. It also means learning on your own by reading books and keeping up with major news events, so that you can maintain the agility necessary to succeed in our increasingly complex modern world. Equally important, it means investing in your personal well-being, so that you can achieve your full potential. This latter type of investment includes obvious things such as eating well and exercising to stay healthy, but it also includes something that too many people seem to have forgotten: successfully managing your personal finances, which is the main topic of this chapter.

Know Your Budget

The single most important step in personal finance is learning to live within your means. Sure, almost everyone wants more money. But in terms of personal happiness, studies show that the amount of money you have is far less important than the degree of control you have over your financial life. People who lose control of their finances tend to suffer from financial stress, which in turn leads to higher rates of divorce or relationship failure, depression, and a variety of other ailments. In other words, if you do not take control of your finances, they will very likely take control of you.

Despite its importance, personal finance is one of those topics that is sadly missing from standard curricula, both in high school and in college. Fortunately, it's not difficult to understand. It's primarily just a matter of paying attention. Know your bank balance, so that you won't overdraw it by accident. Know how much you spend, and on what, so that you can make smart decisions about what to cut or what else you can afford. Know what you are earning in interest on your savings and paying in interest on your loans. Think before you buy, borrow, or invest, so that you don't fall victim to the clichés embodied in the two quotations that begin this chapter.

Easy as all this sounds, you'd be surprised at how many people neglect it. To see if you're one of them, ask yourself a few questions: Do you know the approximate percentages of your after-tax income that you spend on housing, food, and entertainment? Do you know the interest rate on your savings account? If you own a home, do you know the portion of your monthly mortgage payments that is going to principal rather than interest? If you've answered "no" to any of these questions, then you should take some time to create a personal budget.

Creating a monthly budget simply means adding up all your monthly income and then subtracting all your monthly expenses. The result is your monthly net *cash flow*. If your cash flow is positive, then you have money left over to save or invest. If it is negative, you have a problem that will require adjustments. The only trick to this process is making sure you accurately track everything. For example, if you have an expense that you pay once a year (such as college tuition), you can count it as a monthly expense by dividing the annual amount by 12. If you're unsure of what you're spending, try recording everything you spend for a couple of months, then use this record to determine your spending in various categories.

Once you've made your budget, spend a little time studying it to see if you think it can be improved. There are many ways to do this, but a good starting point is to see how your personal spending compares to the national averages shown in Figure 8. For example, if you are spending more than the roughly 33% average on housing, it's good to evaluate why; only then can you decide whether it might make sense to downsize.

Of course, you should never go by averages alone, and there are few strict rules when it comes to finance. The key is to think through everything, in particular the long-term effects of your decisions. As an example, consider a college student struggling to pay tuition and faced with a choice between going deeper into student loan debt and taking an extra job for twenty hours per week. A short-term analysis would probably favor the job. But because more studying and higher grades are likely to lead to greater earnings down the road, the loan might actually be the better long-term plan, since it would leave more time for studying.

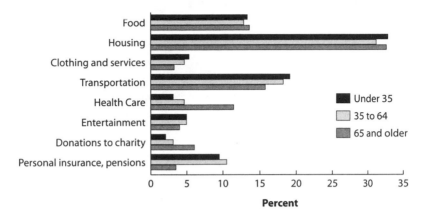

Figure 8. Typical household spending patterns by age group. Data from the U.S. Bureau of Labor Statistics.

Before we move on, it's worth noting that Figure 8 shows only current spending patterns, and these tend to change over time. For example, a century ago the average American family spent more than 40% of its income on food. The fact that food now accounts for only about 13% of spending means much more money is available for other things, which is a major reason we now spend much more on leisure activities than did our great-grandparents. Looking to the future, most economists expect health care to consume a growing percentage of income for all ages, and rising energy costs are likely to mean transportation costs will do the same. If that comes to pass, we'll all have to reduce the percentages of our income that we spend in other areas.

The Changing Value of Money

After budgeting and spending decisions, the biggest issues that most of us face in personal finance are the choices we make for our investments and for any loans we need, including home mortgages and credit cards. We'll therefore devote most of the rest of this chapter to those topics. But to fully appreciate them, we first need to think about the way that money tends to change in value with time. After all, both investments and loans can last for long periods of time, and $100 at some time in the future will probably not mean the same thing that it does today.

At most times, inflation raises wages and prices, which means the value of a dollar declines with time (see Figure 3 on p. 47). Once in a while, prices and wages may go down, which we call *deflation*. Deflation has long been rare in the United States, though it has caused significant problems in a few other countries in recent decades, most notably in Japan, and some economists worry that it could take hold at some point here.

The constantly changing value of the dollar means that we can fairly compare amounts from different times only if they are "adjusted for inflation" (or deflation). To see why, consider gasoline prices. In 1980, for example, the average pump price of gasoline in the United States was about $1.22 per gallon. That sounds remarkably cheap compared to today's prices. However, everything was cheaper back then, so to make a fair comparison we need to adjust the gas price for inflation. When we do that, we find that the 1980 price of $1.22 was equivalent to about $3.23 in 2010, which was actually a bit higher than gas prices most of that year. In other words, while the pump price of gasoline rose significantly between 1980 and 2010, the *real* price—meaning the price in comparison to other prices—dropped a little.

The main tool used to monitor trends in overall prices is the *consumer price index*, or CPI. We'll talk about how it's measured shortly, but first you should understand how it's used. In the United States, the CPI is currently defined so that "100" represents average prices from about mid-1983. In mid-2010, the CPI was about 217. Since 217 is 2.17 times 100, this tells us that prices in 2010 were, on average, about 2.17 times those in mid-1983. For example, an item that cost $1.00 in mid-1983 had a price "adjusted for inflation" of $2.17 in 2010. It's not hard to calculate inflation-adjusted prices,[12] but an even easier way to find them is by using an online inflation calculator, such as the one shown in Figure 9.

---

12. Here's how you do it: Suppose you want to adjust the 1992 price of an item to find its equivalent 2010 price. First, look up the CPI for each of those years; we'll call them CPI_{1992} and CPI_{2010}. Then the equivalent 2010 price is simply the 1992 price times CPI_{2010}/CPI_{1992}. You can use the same procedure to adjust prices between any other two years.

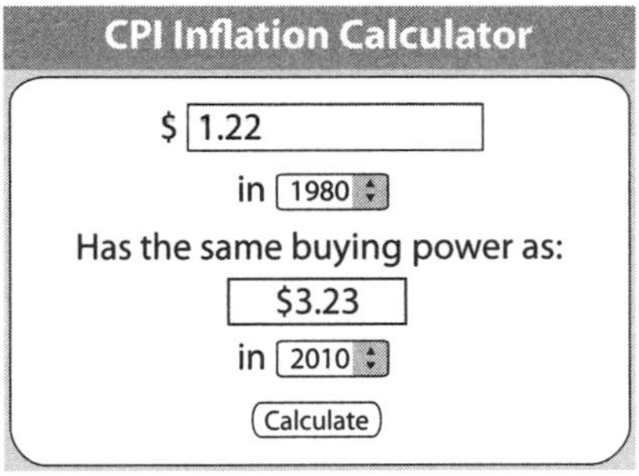

CPI Inflation Calculator

$ [1.22]

in [1980 ↕]

Has the same buying power as:

[$3.23]

in [2010 ↕]

(Calculate)

Figure 9. The inflation calculator on the U.S. Bureau of Labor Statistics Web site.

The CPI may seem a bit abstract, but it has a huge impact on your personal finances. For example, the amount of income tax you pay depends on it, because the government uses it to adjust the thresholds at which different tax rates take effect. Similarly, the government uses the CPI to make annual changes in Social Security payments, with huge implications for those who depend on Social Security for their income. The CPI also has great impact on the private sector, since it is the benchmark used to consider price and wage increases. If a business raises prices by more than the overall rate of inflation (as measured by the CPI), customers are likely to be upset. But if it does not raise wages enough to keep up with inflation, employees are likely to be upset.

That brings us to the key question about the CPI: It's used to represent overall inflation, but does it really measure inflation accurately? To understand this question, we need to look at how the CPI is measured. Each month, the U.S. Bureau of Labor Statistics collects prices for a sample of more than 60,000 items related to consumer purchases, services, and housing. The chosen items are supposed to represent the overall spending patterns of typical consumers, and the CPI essentially tells us how the average price of those 60,000 items today compares to the average price of the 60,000 items measured at times in the past.

You can probably see where problems might come in. The CPI would be a perfect measure of inflation *if* the 60,000 items perfectly represented overall spending patterns and always stayed the same. But since they don't, the CPI can be affected by both random and systematic errors. The large size of the sample (60,000 items) tends to keep the random errors fairly small, so most of the concern is with possible systematic errors. The most obvious potential error comes from changes in "typical" spending patterns. For example, today's CPI includes items such as high-speed Internet access, tablet computers, and smart phones, since many consumers commonly purchase these. However, because those items didn't even exist a decade ago, their prices are in essence being compared to the prices of different items that consumers purchased in the past. Many economists believe that this (along with some other systematic errors) makes the CPI overstate inflation, arguing that a flat CPI would still mean an increase in the average standard of living because we are able to buy ever more sophisticated products. Of course, you can find other economists who argue the reverse.

This is a high-stakes debate. Suppose the CPI is overstating inflation. In that case, the annual adjustments the government makes to such things as tax brackets and Social Security payments are too large. Fixing this overstatement would therefore raise your taxes and/or lower your Social Security payments, while also lowering the government's deficit by $100 billion or more per year. The situation would be reversed if the CPI understates inflation. Personally, I find the argument for overstatement more persuasive, which puts me in agreement with most economists I've talked to. I therefore hope we'll find the national will to confront this issue, even though it will have some direct impacts on all of us.

But for this chapter, the key point is simply that prices tend to rise with time. Therefore, if you want to keep up, you need to find a way to make your money grow with time as well. And that brings us to the topic of compound interest.

Compound Interest

On July 18, 1461, King Edward IV of England borrowed the modern U.S. equivalent of $384 from New College of Oxford. The king soon paid back $160, but never repaid the remaining $224. The debt was forgotten for 535 years. Upon its rediscovery in 1996, a New College administrator wrote to the queen of England asking for repayment, with interest. Assuming an interest rate of 4% per year, which he figured was about the right average for the period, he calculated that the college was owed $290 billion. In a stroke of generosity, he offered to settle for an interest rate of only 2% per year, in which case the college was owed only $8.9 million.

Unfortunately for the college, there was no clear record of a promise to repay the debt with interest, and the queen presumably did not take the letter too seriously. Nevertheless, this example shows three key facts about compound interest. First, it shows that the same ideas of interest apply to both investment and debt. That is, one person's debt is another person's (or bank's) interest-bearing account. Second, it shows what is sometimes called the "power of compounding": the remarkable way that money can grow when interest continues to accumulate year after year (at least if the interest is actually paid). Third, it shows the extreme sensitivity of earnings to the interest rate. If you didn't know better, you might guess that doubling the interest rate from 2% to 4% would also double the total value of the debt. In fact, the $290 billion that accumulates over 535 years at 4% is more than 32,000 times as much as the $8.9 million that accumulates at 2%. Most of us don't invest quite that long term, but small differences in interest rate can still amount to a lot of money.

The basic principle of compound interest is easy to understand. Imagine that you open a bank account with $1,000, and that you will receive interest at an annual percentage rate, or APR, of 5%. To keep things simple, let's assume the interest is paid just once a year. At the end of the first year, you'll be paid 5% of $1,000, which is $50, bringing your balance to $1,050. At the end of the second year, you'll be paid 5% of this new balance, which is 0.05 × $1,050 = $52.50; that brings your balance up to $1,102.50. In the third year, you'll receive 5% of this new amount, which is $55.13. And so on. Notice that even though the interest rate stays fixed at 5%, the dollar amount of the interest rises each year, because you are receiving interest on the previous interest as well as on the original principal. It is this idea of "interest on interest" that financial folks refer to as *compounding*.

The difference between APR and APY. Suppose that the interest is paid, or compounded, twice a year instead of once a year. In that case, you would receive half of the 5%, or 2.5%, at the end of six months, and the other 2.5% at the end of the year. But notice what happens: The first payment is 2.5% of your starting balance of $1,000, or $25. Because that brings your balance to $1,025, the second payment is .025 × $1,025 = $25.63, making your year-end balance $1,050.63. By having the interest split into two payments during the year, you end up slightly ahead. In this particular case, rather than growing by exactly the APR of 5% each year, your money grows by 5.063% each year; this is what banks usually refer to as the *annual percentage yield*, or APY. It's probably clear that more compounding periods mean a higher APY. That is, quarterly compounding builds your balance faster than twice-a-year compounding, and monthly compounding builds it faster than that. Banks usually tell you both the APR and the APY, but the APY is the more important number, because it tells you the actual percentage by which your balance will grow in a year.

It might be tempting to think that if you really had a lot of compounding periods—say, a trillion each second—you could get the APY to be a lot higher than the APR. In reality, the added benefit gets smaller and smaller as you add more compounding periods, so that the maximum possible APY has a theoretical limit that isn't all that different from the APR. For example, for an APR of 5%, the theoretical limit on the APY is about 5.127%.[13] Some banks pay at this theoretical limit, but most compound daily, which gives a very slightly lower APY.

Effects of today's low interest rates. I've been teaching and writing about compound interest for more than two decades, and during most of that period I was able to impress students with the incredible power of compounding by giving them examples like the ones above, with interest rates of a few percent or more. However, since the beginning of the recent recession, the rates that you can earn on bank accounts have plummeted to near zero (and sometimes to zero).

Besides making it difficult to keep pace with inflation, these low rates have had devastating effects on retirees who hoped to live off interest. For example, consider a retired couple who managed to save $500,000 for retirement. At an interest rate of 5%, they would earn $25,000 per year in interest, and could use this money to live on without reducing their $500,000 principal at all. But at the 0.05% that my own credit union account has recently offered, the interest would be only $250 per year. If the couple needed $25,000 to live, they would have to draw down their principal by about this amount each year,

---

13. Compounding as often as is theoretically possible is called *continuous* compounding, and it generates an APY that you can calculate by raising the special number *e* (which is approximately 2.71828 and is found on any scientific calculator) to the APR power and subtracting 1.

in which case their retirement account would be empty in only about twenty years, likely leaving them dependent on Social Security. We'll talk more about the future of Social Security in Chapter 6, but you can see why it is an issue that draws highly emotional responses.

The compound interest formula. That's pretty much all there is to compound interest, and there are lots of online calculators that allow you to compute how your balance will change with time, depending on the APR and the number of compounding periods each year. However, it's so easy to come up with a simple formula that I can't resist showing it to you. Look back at our example of 5% interest computed once a year. Notice that, each year, the new balance is simply the prior balance multiplied by 1.05. For example, since we started with $1,000, the balance after the first year was $1,000 × 1.05, or $1,050. To find the balance after the second year, we multiply that amount by 1.05, giving us ($1,000 × 1.05) × 1.05, which is the same as $1,000 × 1.05^2; after the third year, it is the second year's balance times 1.05 again, or $1,000 × 1.05^3. (Recall that powers tell us how many times to multiply a number by itself, so 1.05^3 means 1.05 × 1.05 × 1.05.) We can generalize this idea for any starting principal P and any interest rate APR paid for Y years; your balance at the end of the Y years will be $P \times (1+APR)^Y$.

Just to show how easy this formula is to use, let's look back at the New College case. The starting principal was $224, so if we assume an annual percentage rate of 4%, or 0.04, and a time period of $Y = 535$ years, then the final balance would be $224 \times (1+.04)^{535}$. You can use a calculator to confirm that $1.04^{535} = 1,296,691,085$, and that multiplying this by $224 gives just about $290 billion. For interest compounded more than once a year, you can use the same formula simply by replacing the APR with the APY.

Investment Options

Even at their higher historical values, bank accounts have rarely paid interest rates high enough to beat inflation by much. For that reason, most people have sought to do better by trying other investments. Although there are far too many ways to invest for us to cover them all here, we can look at a few general principles of investing.

The four most common general types of investment are stocks, bonds, cash (which includes bank deposits and U.S. Treasury bills), and real estate. In general, we can evaluate these investments in terms of three major considerations: liquidity, risk, and return.

Liquidity refers to the ease with which you can withdraw money. An investment from which you can withdraw money easily, such as an ordinary bank account, is said to be liquid. The liquidity of an investment like real estate is much lower, because real estate can be difficult to sell.

The meaning of *risk* is fairly obvious. Safe investments are those in which you are unlikely to lose the money you invest; examples include federally insured bank accounts and U.S. Treasury bills (though this perception of safety might change if the government were to default on its debt). Stocks and bonds are much riskier because they can drop in value, in which case you may lose part or all of the money you invest in them.

The annual *return* on an investment is its percentage change in value during a particular year; it is essentially equivalent to the APR for a bank account, except that it may vary greatly with time, and it can be negative as well as positive. In general, low-risk investments offer relatively low returns, while high-risk investments offer the prospects of higher returns—along with the possibility of loss.

Investing Smartly

The most difficult part of investing is finding an appropriate balance between risk and return. Indeed, it is so difficult that even so-called financial wizards failed miserably at it during the past few years, bringing down long-established institutions and nearly taking the entire economy down with them. Given that track record of professionals, you may wonder how you can guarantee your own investment success, and I'll give you a simple if unsatisfying answer: You can't, because there are no guarantees when it comes to investments. Moreover, as the quote at the beginning of the chapter tells us—and as most of Bernie Madoff's investors found out the hard way—"if it sounds too good to be true, it probably is."

Nevertheless, there are some obvious differences between smart investors and those who come close to destroying the economy. To start with, *don't be greedy*. If you look for an outsize return, it automatically comes with outsize risk. *Invest for the long term*. Don't buy a house because you think you can make a quick profit on it; buy it because you want to live there. And follow the old dictum, "Don't put all your eggs in one basket." That way, if one investment goes wrong, you can hope that your others might at least partially make up for it.

It's also worthwhile to look at historical trends, as long as you don't forget the crucial caveat that "past performance is no guarantee of future success." These trends show that over the longest term, stocks have been the best investment. For example, imagine that in the year 1900, your great-great-grandmother had invested $100 each in stocks, bonds, and cash. If she was smart enough to choose a financial firm that actively managed her portfolios so that they tracked the averages for each of those investments, then here is approximately what you would have inherited as of 2010: the cash would be worth about $300, the bonds would be worth about $950, and the stocks would be worth about $54,000.

Of course, before you rush out and invest all your money in stock funds, remember that you may need access to your money in less than 110 years, and there have been long stretches in which stock investments would have lost money. For example, if you had invested in a mutual fund that tracked the Dow Jones Industrial Average (DJIA) just before the crash of 1929, it would have taken about twenty-five years for the investment to recover its initial value. If you invested in 1999, a decade later your portfolio was worth nearly a third less. And if you were unfortunate enough to invest in Japan's Nikkei stock index when it hit its high in 1989, your investment in 2011 was worth some 70% less than what you put in.

Understanding Stocks and Bonds

These days, almost everyone with a retirement plan owns some stocks and bonds, but not everyone understands exactly what these are. As with most financial instruments, the details can be complex, but the basic ideas are simple.

Stock (also called *equity*) gives you partial ownership of a company. If the company has issued a total of 1 billion shares, and you buy one share, then you own one-billionth of the company. You can therefore find the total value of a company, called its "market capitalization," or "market cap" for short, by multiplying its stock price by the total number of shares in the company (the "shares outstanding"). Figure 10 shows a sample stock quote for Microsoft. At the time of the quote, the share price (called "last," because it was the last price available) was $19.76 and there were 8,899.72 million (or 8.89972 billion) shares outstanding, so Microsoft's market capitalization at that moment was $19.76 × 8.89972 billion ≈ $176 billion; this is about $1 billion lower than the market cap shown in the figure, because the stock quote shows the company's market cap at the end of the previous day, when the stock price was a slightly higher $19.89 (shown as "prior day's close"). Be sure to note that the stock price by itself does not tell you anything about the company's total value, because companies can have vastly different numbers of shares outstanding. In fact, companies with high share prices sometimes decide to "split" their shares; for example, in a 2-for-1 split the company turns every share into two shares, which means the share price falls by half at the same time, so that the total value of the company is unchanged. (Similarly, companies with low prices may do a "reverse split" to make the share price higher.)

If you are going to invest in stocks, you should make sure you understand all the data in stock quotes such as the one in Figure 10. Along the left the quote shows basic data on the stock price, including its current ("last") price per share, the opening and high and low prices for the day, the high and low prices for the past 52 weeks, and the change in price since the prior day both in dollars and as a percentage. In the middle it shows the total number of shares traded for the day ("volume"). To the right, you see the market capitalization at the end of the prior day (at top) and the shares outstanding (at bottom). The "P/E ratio" is the share price divided by the company's earnings, or profit, per share over the past year. Historically, the average P/E ratio has been around 12, meaning that a company that earned $1 per share in profits would have stock selling for about $12 per share. If the P/E ratio is much higher than this, it usually means that investors are speculating on the stock in the belief that the company's future profits will be much higher than its present profits; conversely, a low P/E ratio may indicate that investors think profits are likely to go down in the future, making the company less valuable.

There are two basic ways to earn money from stock investments. First, some companies pay a dividend, which is like interest because it gives you a direct return on your investment. Dividends are typically paid quarterly as

Microsoft Corp. (MSFT)			NASDAQ	Market Cap ($millions)	$177,015.40
Comprehensive Quote:			05/13/09 03:05 PM EDT		
Last 19.76	Change −0.13	% Change −0.64%	Volume 33,247,937	P/E Ratio	11.45
Open 19.92	High 20.00	Low 19.67	Prior Day's Volume 71,966,786	Dividend (latest quarter)	$0.13
52-Week High 30.53	52-Week Low 14.87		Prior Day's Close 19.89	Dividend Yield	2.61%
				Shares Outstanding (millions)	8,899.72

Figure 10. A typical online stock quote.

some amount per share. For example, Figure 10 shows a dividend of 13¢ per share, so if you own 100 shares, your total dividend paid at the end of the last quarter was $13.00. The "dividend yield" is the APY that you would be receiving on your investment if both the stock price and the dividend per share stayed unchanged. The second way to earn money on stocks is by selling them for a higher price than you buy them for, which means stock buyers are always shopping for companies that they think will increase in value. Of course, the company may also go down in value, in which case you could end up losing some or all of the money you invested in it. It's also worth remembering that when one person thinks a stock is a bargain and decides to buy it, the shares are available at that price only because others think it's overpriced and therefore are willing to sell.

Bonds are a form of debt in which the bond issuer is indebted to you. If you buy a $1,000 bond issued by MegaTech Corporation, MegaTech is promising to pay back your $1,000 with interest. The $1,000 is the *face value* of the bond, and the promised interest rate is called the *coupon rate*. The major difference between investing in a bond and investing in cash (such as a bank account or Treasury bill) is risk: If MegaTech goes bankrupt, you'll lose your $1,000 and never see the interest. For that reason, the interest rate on a bond depends on the financial strength of the company (or other entity) that issues it. Strong companies can issue bonds at relatively low interest rates, but no one will buy bonds from a weak company unless they promise a relatively high interest rate to make up for the associated risk. "Junk bonds" are high-interest-rate bonds from companies considered extremely risky; they work out great if the company actually pays you back, but they are rated "junk" because there's a high probability that you'll lose your entire investment.

Notice that, because the financial strength of a bond issuer can change, the value of bonds can also change. For example, suppose your $1,000 Mega-Tech bond has a coupon rate of 8%. That's a high interest rate these days, presumably reflecting the risk that MegaTech will fail. But if MegaTech becomes a stronger company, the risk goes down, and someone might then be willing to buy your 8% bond for more than the face value of $1,000 (which is known as paying a *premium* for the bond). Conversely, if MegaTech becomes weaker and you want to sell your bond, you'll probably have to accept less than the face value of $1,000 (a *discount* on the bond price).

The inherent risks mean that you should never buy stocks or bonds for an individual company unless you carefully research the company's financial health. That obviously takes a lot of time, which is why most investors steer toward mutual funds. A mutual fund is a pool of money managed by a professional who has time for all the research, taking the burden off you. Today, you can choose from many thousands of mutual funds, each with particular strategies. Some invest only in stocks for particular types of companies (such as energy companies), others invest only in bonds, and still others hold a mix of stocks, bonds, and cash. *Index funds* are managed so that their value tracks the performance of an index such as the DJIA or the S&P 500. Many financial advisors recommend index funds because they essentially track a broad part of the economy, taking out most of the guesswork that otherwise goes with selecting companies or mutual funds. One warning: Mutual funds virtually always charge some sort of fee for their services, and high fees can dramatically eat away at your investment. Be sure to shop carefully for mutual funds with both good track records and low fees.

Borrowing Money

It would be nice if we all had enough money so that we only had to worry about investing. In reality, nearly all of us need to borrow money, whether to buy a house or a car, to go to college, or just to get through between paychecks. Mathematically, loans are nothing more than bank accounts in reverse; instead of depositing money and collecting interest from a bank, you borrow money and pay interest. However, lenders usually expect you to make payments on some schedule (such as once a month), which adds some subtleties that we can best understand through examples.

Suppose that you borrow $1,000 at an annual interest rate (APR) of 12%, and that you are required to make monthly payments. The 12% annual rate means that your monthly interest rate is 1%. Because 1% of $1,000 is $10, your first monthly payment will require $10 in interest. If you paid only this $10 in interest, then the amount you owe, called your loan *principal*, would stay fixed at $1,000. In that case, you'd owe the same $10 in interest the next month, and if you continued to pay only the interest, your loan principal would never go down.

If you actually want to pay off the loan, you'll need to pay some of the principal in addition to the interest each month. For example, if you paid $200 toward the principal each month, then it would take you five months to pay off the $1,000 loan. But notice what happens to the interest in this case. The first month, your interest would be the $10 we found earlier. However, by paying $200 toward the principal, you would reduce your loan principal to $800. Therefore, at the end of the second month, your interest payment would be 1% of $800, or $8. Paying another $200 toward principal would leave a loan balance of $600, so the third month's interest would be 1% of $600, or $6. Because your interest payment gets smaller each month while your principal payments stay the same, your total payment (principal plus interest) gets gradually smaller.

The above strategy is a perfectly good one for paying off a loan. However, most people prefer to have a loan payment that stays the same each month rather than one that changes with time, because that makes it easier to stick with a budget. Such loans are called *installment* (or *amortized*) loans. To understand how installment loan payments work, let's continue our above example. We already know that you'd need to pay $200 per month to pay off the principal in five months, even without any interest at all. We also know that the first month's interest is $10. Therefore, if you want equal monthly payments, the payment amount is going to be somewhere between $200 and $210; an exact calculation shows that the monthly payments would be about $206. Here's how it works out: The first month, interest takes $10 out of the $206, with the other $196 going toward the principal. The reduced principal means your interest payment will drop to just over $8 the second month, and since you are paying the same $206, the amount going toward principal will increase to nearly $198. The interest payment falls further with each passing month, which means corresponding increases in the portion of the $206 going to principal. In other words, fixed monthly payments gradually go more and more toward principal, and less and less toward interest.

In case you're wondering, I found the exact payment (which is actually $206.04) using a formula that is only slightly more complex than the compound interest formula. However, there are so many online loan payment calculators that I won't bother showing you the formula here. Simply enter the starting principal, the interest rate, and the loan term into one of these calculators, and it will pop out your exact monthly payments. Even better, many of these calculators will generate an *amortization schedule*—a table that shows you the portion of the payments going to interest and the portion going to principal each month or each year, along with a summary of the total interest paid over the life of the loan.

When you are considering a loan, it's worth spending some time to evaluate the results of different scenarios. For example, you'll often have a choice between a longer-term loan at a higher interest rate or a shorter-term loan at a lower interest rate. Use an online calculator to find your payments and amortization schedule in each case. The results can be very surprising; much like compound interest, a high interest rate and a long term can lead to very high total interest payments. Consider a home loan for $150,000. At an interest rate of 6% and a term of 15 years, your monthly payments will be about $1,266 and you'll pay a total of about $78,000 in interest during the 15 years. At a slightly higher interest rate of 6.5% and a loan term of 30 years, the monthly payments fall to $948, but the total interest over the 30 years rises to more than $191,000. Moreover, for a significant portion of the 30-year term, the vast majority of your monthly payment goes toward interest. From this perspective, the shorter-term loan is always better. Of course, you also need to consider what monthly payments you can afford; after all, the entire point of a loan is to spread out your costs over time.

You should also consider tax implications, inflation, and your investment options. For example, if the mortgage tax deduction happens to make your effective interest rate lower than the inflation rate, then you may be better off (at least in purely financial terms) stretching out your loan payments for as long as possible. Similarly, if you have investment options that you think are likely to have a higher annual return than the interest rate you're paying, you may want to keep your monthly payments lower. Many financial advisors put great weight on these types of considerations, though personally, I find the psychological comfort of knowing that I might someday be debt-free to be worth a lot more than most of these small financial benefits. That's why I generally recommend opting for the lower-interest, shorter-term loan, as long as you're confident that you can afford the higher monthly payments.

Choosing a Loan

Interest rate and loan term determine your monthly payments on any particular loan amount, but they are not the only important factors in choosing a loan. Here are four others that are particularly important to loans such as home mortgages, auto loans, and student loans.

First, you may have a choice between a fixed-rate loan and a variable-rate loan, such as an adjustable-rate mortgage (ARM). *Fixed* means that the interest rate, and therefore your monthly payments, will not change for the life of the loan. That obviously makes it much easier for you to plan future budgets than a loan for which your monthly payments could dramatically increase. The potential advantage to an adjustable-rate loan is that it usually comes with a lower interest rate, at least initially. In principle, then, the trade-off is pretty simple: If you think you'll be keeping your loan for a long time, stick with a fixed rate; if you expect to pay off your loan early (for example, by selling the home or car on which you have the loan), before an adjustable rate can tick higher, then the adjustable loan may work for you. Be careful, however, as many adjustable-rate loans start with an artificially low "teaser rate" that lasts only a short time. These low rates can make your monthly payments seem easy to afford at first, but the payments can soar beyond your budget once the rates tick upward. My advice: Unless you have some very well-thought-out rationale for an adjustable-rate loan, stick with a fixed rate.

Second, be sure you understand all the fees and closing costs that will be charged. These can be very confusing, and lenders often try to hide them or pretend they are something else. For example, many mortgages require an "origination fee" that is typically 1% of the total loan amount. Easy enough to understand, but then they also charge "points"; for example, 1.5 points means 1.5% of the loan amount. What's the difference between an origination fee and points? Nothing, really. You may even be offered the option to pay extra points in return for a lower interest rate. To decide if this is a good idea, you need to figure out how long it will take you to recoup the cost of the extra points through lower payments, and then do it only if you are virtually certain you'll own the house at least that long.

Other fees can vary a great deal from one lending institution to another, which is why you should always shop around for loans. Fees on student loans have been particularly notorious, though this should in principle get better thanks to recent legislation. Perhaps the worst possible loans in terms of fees are the so-called payday loans, which allow you to borrow a small sum of money with your coming paycheck as collateral. A typical payday loan might give a person $500 for two weeks for a fee of $75. If you work it out as though it were a normal loan with compound interest, you'll find this translates to an effective annual interest rate of 3,685%. Clearly, it's a very bad deal for the borrower, though a very profitable one for the lender.

A third crucial factor is the fine print. This issue can be particularly difficult to deal with, because the fine print in most loans goes on for pages and pages in language that even most lawyers don't fully understand. Nevertheless, you should still make the effort to read through the loan contract in search of anything that might make the loan more expensive than it seems on the surface. In particular, make sure you check that the loan contract does not contain any prepayment penalties, since you should always retain the right to pay the loan off early or to refinance it at a better interest rate. Be sure to look for these things yourself—don't just trust what the lender tells you. My wife once was about to take out a home loan for which the lender assured her multiple times that there would be no prepayment penalty. But when she went to sign the papers, she read them carefully enough to discover that a prepayment penalty was in there, and she refused to sign until they removed it. If she had not been so diligent in reading the fine print, she would have ended up signing a loan that could have cost her a lot of money when she sold the house later.

Fourth, many loans require a down payment. While a down payment can seem onerous, it's usually a pretty good idea. Often, you can get a better interest rate if you make a larger down payment. Moreover, making a down payment is a good reality check. If you've already saved 20% of a home's price to use as a down payment, it bodes well for your ability to budget for your loan payments down the road. Again, the recent mortgage crisis provides a cautionary lesson. Many people were able to buy homes with little or no money down. As a result, they had less understanding of whether the homes were truly within their budgets, and when home prices dropped, they had no equity in the homes at all.

The bottom line is that taking out any loan is going to require you to apply nearly all the different types of mathematical thinking that we've discussed so far in this book. Remember that lenders almost always want to find some way to get you to borrow, because that's how they make money. So regardless of what a lender might tell you, it's critical that *you* feel confident that you will be able to afford your loan payments throughout the life of the loan.

Refinancing a Loan

Once you have a loan, you may be tempted to refinance it. This can be a good idea in some circumstances, particularly if it lowers your interest rate, but far too many people refinance without thinking through the costs and benefits. Two particular concerns jump out. The first is all those fees and closing costs, which you are going to have to pay all over again. For example, suppose you can refinance a loan to a lower interest rate that reduces your monthly payments from $1,000 to $950. The $50-per-month savings sounds good, but if you have to pay $2,000 in up-front fees, then it will take you $2,000 ÷ $50 = 40 months, or more than 3 years, to actually save any money. As a rule of thumb, you should not refinance a loan if it will take more than about two to three years to recoup the up-front fees, unless you are certain you will be holding the loan much longer. Second, and more often forgotten, is that refinancing "resets the clock" on a loan. For example, suppose you have been paying off a ten-year student loan for four years. If you keep this loan, you will pay it off in six more years. But if you refinance with a new ten-year loan, you will have payments for ten years starting now. So even if the new loan reduces your monthly payments substantially, it may not be worth it.

Mortgage Follies

We've discussed a few of the pitfalls that helped create the mortgage crisis, but the real problems went far deeper. Although we can hope that the era of these shenanigans is over, lenders in the mid-2000s convinced a lot of people to take out some truly insane loans. One example was "interest-only" loans, which allowed you to pay only the interest for some set period of time, with nothing going toward principal. This meant that you would eventually need to refinance, so you'd be stuck if the value of your home had dropped by that time (as it had for many people); after all, no one's going to give you a $200,000 loan for a house that's worth only $150,000. We could list many more such insanities, but the bottom line remains simple: Thousands of lenders promised deals that were too good to be true, and millions of people believed them.

Debit and Credit Cards

Debit and credit cards look nearly identical, but they are quite different. A debit card is like a check, because it takes money directly out of your bank account. In contrast, every time you use a credit card you are taking out a loan. This makes credit card lending a big business. On average, American households carry a credit card balance of more than $9,000; if you exclude households without credit cards, the average balance for those with them is about $16,000.[14]

Credit cards can come in very handy, and the itemized statements you receive can help you keep track of your budget. Moreover, if you pay off your balance in full each month and avoid interest payments and fees, then the card issuer is effectively giving you a free short-term loan for your purchases; you may also receive rewards such as airline miles or cash back. However, for those who don't pay off their full balance each month, credit cards are generally a very poor choice.

---

14. These balances are means found by dividing total U.S. credit card debt by number of households. Medians are significantly lower, since households with large debt skew the mean upward.

The major problems with credit cards arise from two important differences between the way their loans work and the way installment loans work. First, unlike installment loans, which have monthly payments designed to get the loans paid back, credit cards generally require only minimum payments designed to keep you in debt for as long as possible. In many cases, these payments cover only the interest, so you would *never* pay off your credit card balance if you made only the minimum payment each month. Why would card companies set minimum payments that don't pay off the principal? The answer is the second major difference between credit cards and other types of loans: With rare exceptions, the interest rates on credit cards are exorbitantly high. For example, as I write this book, the average home loan interest rate is about 5%, while the average credit card interest rate is over 14%, and many cards charge 20% or more. Credit card companies therefore profit far more by keeping you in debt than by encouraging you to pay off your loans.

The trick to credit cards, then, is finding a way to get their upsides without their downsides. This is actually pretty easy to do, if you are disciplined. For example, try not to use more than one card, since a single card makes it easier to keep track of your total balance and your budget. Don't use your credit card to get cash (use your ATM card instead), because you'll inevitably be charged a ridiculously high fee for the privilege. Make sure you are never late with payments, since that also will lead to huge fees. Most important, make sure your credit card has a "grace period" that allows you to avoid interest if you pay off your balance each month, and be sure to do that. In other words, if you need a loan, find a way to get one with a more reasonable interest rate and payment plan than you can get from a credit card; if you own a home, a home equity line can be a particularly good option. Finally, if you find yourself unable to pay off your credit card debt, consult a financial advisor right away; a good place to start is with the National Foundation for Credit Counseling (www .nfcc.org).

Math for Life

When I talk to people about the material that we've covered in this chapter, one of the most common reactions that I get is, "But doesn't everyone already know all this stuff?" After all, we've been discussing ideas that arise in nearly every major financial decision that any of us ever make. My response is always simple: Just look around at the economic mess we've made as a nation, and it's pretty clear that a lot of people don't really understand basic finance.

Of course, this begs the question of why not, and to answer that one, ask yourself (or someone else for whom this chapter was all review) where you first learned about it. Unless you were an economics or business major, the answer is unlikely to be "in school." More likely, you learned it from reading the newspaper, or in the course of your own personal financial experiences; sometimes, you may have learned it the hard way.

The sad fact is that our schools have been just as negligent in teaching basic financial literacy as they have been at teaching how to think with numbers, or statistics, or any of the other topics we'll cover in this book. Indeed, I've spoken to friends with PhDs in mathematics or science who didn't know the meaning of the consumer price index, or the difference between stocks and bonds, or what lenders mean by "points" on a home mortgage.

So in closing, I want to return to an idea I've mentioned before: We can't expect people to know things that they've never been taught. Given the importance of financial literacy, it's critical that we start teaching it in school. I hope you'll join in the growing movement to get high schools and colleges to incorporate financial literacy into their curricula. Better yet, don't do it in isolation from other "math for life" topics; instead, introduce it as part of a general course in quantitative reasoning, and then hammer it home by making sure that it is integrated throughout the school curriculum.

5

Understanding Taxes

In this world, nothing is certain but death and taxes.
— **Benjamin Franklin**

The hardest thing in the world to understand is the income tax.
— **Albert Einstein**

Question: Based on statistical averages, rank the following individuals according to the percentage of income they pay in taxes to the federal government, from the lowest percentage to the highest.

Answer choices:
a. A fast-food worker earning $8 per hour
b. A teacher earning $52,000 per year
c. A self-employed businesswoman earning $100,000 per year
d. A midlevel executive earning $200,000 per year
e. A billionaire investor

Before we discuss this answer, note that this type of *ranking task* is becoming increasingly popular in education, for good reason. For one thing, it requires more thought than a standard multiple-choice question, because you have to consider all of the options, as opposed to just looking for a single correct one. It also reduces the probability of a correct answer through guessing. For a five-choice multiple-choice question, a pure guess will give the right answer one out of five times. In contrast, a five-option ranking task means you have five choices just for the first position in the list. Once you pick a choice for that position, you'll have four choices remaining for the second position, then three for the third and two for the fourth, before you are left with just one choice for the last position. Therefore, the total number of possible ranking orders is $5 \times 4 \times 3 \times 2 \times 1 = 120$, which means you have only a 1 in 120 chance of getting the correct answer with a pure guess. Of course, as with any new trend, there's a danger of overuse, so teachers should remember that not all concepts lend themselves to ranking questions, and that there's a trade-off between their advantages and the fact that they take longer to complete, which means you can't test as many concepts with them as you can with multiple-choice questions in a timed test. It's also much more difficult to write good ranking tasks than good multiple-choice questions.

Anyway, take your best shot at our current question. If you put A in the first spot, you're off to a good start. The $8-per-hour fast-food worker almost certainly pays the lowest percentage of income in federal taxes. In fact, depending on the specific circumstances, he or she might pay nothing at all, and could even be *receiving* money from the federal government through the earned income tax credit.

The rest of the positions are more difficult. You might guess that the percentage would go up in income order, but taxes are much more complicated than that. We'll spend much of this chapter discussing the details, but for now let me just give you the basic answer, which is A-E-B-D-C. If you're like most people, your next thought is probably along the lines of "You've got to be kidding."

But I'm not. After the poverty-level fast-food worker, the billionaire investor pays the smallest percentage of income in taxes. The reason is that investment income, called *capital gains*, is taxed at a much lower rate than *earned income* from a job. In most cases, investment income is taxed at a flat rate of 15% (under tax law as of 2011).

The teacher comes next. Again, the details will vary greatly with personal circumstances, but the teacher's relatively low income will probably yield an overall income tax rate between about 10% and 15%. Wait, you say: that's lower than the billionaire's, not higher. But note the key words "income tax." In addition to income tax, earned income (but not capital gains) is subject to FICA taxes, which pay for Social Security and Medicare. (FICA stands for the Federal Insurance Contributions Act.) The teacher will be paying a flat 7.65% for FICA, and adding that to his or her income tax yields a higher overall tax rate than the billionaire's.

That leaves us with the self-employed businesswoman (C) and the executive who earns twice as much (D). Surely, you might think, the higher-earning executive must pay more. But no . . . Again, FICA creates a surprising result. The executive will almost certainly pay a higher rate of income tax, though not by more than about 5 to 10 percentage points. While this might sound like a significant difference (it could be the difference between 15% and 25%), it's easily overwhelmed by the difference in FICA payments. Although most people don't realize it, the 7.65% that gets pulled out of your paycheck for FICA must be matched by your employer. That means the self-employed businesswoman must pay both her personal share and the employer share, for a total rate of 15.3%. Moreover, the Social Security portion of the FICA payment, which is most of it, goes away above a certain income level. For 2011, that level was $106,800. Therefore, the $200,000 executive pays the 7.65% FICA rate on only about the first half of his income, for an effective rate of less than 4% on his full income. In contrast, the businesswoman is just short of the cutoff, which means she pays the 15.3% self-employed FICA tax on her entire income.

Of course, all this is based on statistical averages; like dietary supplements, taxes come with the caveat that "individual results may vary." Your personal situation matters. Married people pay at a different rate than singles; having more children at home usually reduces your tax bill (though it increases your other bills). Most critically, a long list of special tax breaks, including numerous tax credits and deductions, means that two people with exactly the same income can have vastly different tax bills. In recent years, this has been further exacerbated by the *alternative minimum tax*, or AMT, which was originally designed to prevent the rich from using so many loopholes that they ended up with no tax bill, but which now ensnares millions of taxpayers of more modest means.

You've probably now guessed why I've included an entire chapter on taxes in this book, but just in case, let me be clear about my two main reasons. First, if you're like most people, taxes are either your largest or second-largest (after housing) personal expenditure, which means you can't possibly do the kind of budgeting we discussed in the last chapter unless you understand what you are paying in taxes. Second, I challenge anyone of any political persuasion to consider the answer to our ranking task (not to mention the AMT and other complications) and claim that our current system makes sense and meets reasonable standards of fairness. Our tax system needs changes, and while coming to agreement on the changes is going to be very difficult, understanding what we're dealing with is a clear first step.

Types of Tax

The ranking task gave us the opportunity to talk a bit about the two major types of federal taxes: income tax and FICA. But we pay many other taxes as well, including state income taxes, sales taxes, property taxes, gasoline taxes, and "sin taxes" on products like tobacco and alcohol. With such great variety in taxation, it's helpful to do some categorization that will lend order to the chaos. Although there are lots of subtleties and variations, most taxes fall into one of three basic categories: income taxes, sales taxes, and property taxes.

An *income tax* is any tax that is calculated based on some percentage of income. That means FICA is essentially an income tax, even though we don't usually call it that. Most of us pay at least three different types of income tax: federal income tax, state income tax, and FICA.

Sales taxes are sometimes called *consumption taxes*, because they are taxes on what we spend (or "consume") rather than on what we earn. Nearly all purchases are subject to state and local sales taxes. Gasoline taxes and sin taxes are also sales taxes, since they are based on what we buy.

Another type of sales tax, and one that you are likely to be hearing a lot more about, is a *value-added tax*, or VAT. Although we do not currently have a VAT in the United States, it's very common elsewhere; more than 100 countries have a VAT, including Canada, Mexico, and all the countries of the European Union. For consumers, the VAT works like any other sales tax, in that it is included in the final price you pay for your purchases. For example, if the total VAT for a loaf of bread amounts to 10%, then you'll pay 10% more than you would without a VAT, just as you would if there were a 10% sales tax. The reason the VAT gets a special name is that it works a little differently on the business end. For the loaf of bread, let's assume that it reached the supermarket through a supply chain encompassing the farmer who grew the grain, the baker who turned the grain into bread, and the distributor who delivered the bread to the supermarket. With a VAT, each business in the supply chain must add some tax when it sells to the next business, and the 10% tax that you pay at the supermarket is actually the sum of all the taxes collected and paid to the government by all the businesses. The name "value-added" comes from the idea that each business in the supply chain pays a tax based on its contribution to the value of the final sale.

Finally, *property taxes* are based on the estimated value of property that you own. Most of us pay two types of property tax: a tax on the value of real estate that we own, and a tax on the value of the cars that we own.

Flat, Regressive, and Progressive Taxes

There are pros and cons to any particular type of tax, but one of the most fundamental ways in which taxes are evaluated is by whether they are flat, regressive, or progressive.

A *flat tax* is one with the same tax rate for everyone. FICA is a flat tax, at least up to the point where it cuts off for high incomes, because everyone pays the same 7.65% rate (or double that if you're self-employed). A few states have a flat income tax, in which everyone pays the same percentage of income, at least above some minimum threshold. For example, as of 2010, Colorado had a flat income tax rate of 4.63%, and Indiana had a flat rate of 3.4%. Property taxes are usually also flat, at least in principle, because the tax is a simple percentage of the assessed value of your property.

A *regressive* tax is one that takes a higher percentage from poorer people than from wealthier people. Sales taxes are an example, particularly when they are charged only on goods and not on services (as is commonly the case in the United States), because poorer people generally spend a much larger proportion of their income on goods than richer people. As a result, poorer people pay a much higher percentage of their overall incomes in sales taxes, even though the sales tax rate on any particular item is the same for everyone. In principle, the regressive nature of sales taxes can be offset by exempting certain items (such as groceries), taxing "luxury" items at higher rates, or providing rebates. Those who advocate a VAT (or other consumption tax) for the United States usually suggest these types of tweaks as a way of making sure it does not hit poor people the hardest.

The opposite of a regressive tax is a *progressive* tax, in which richer people pay a higher percentage rate than poorer people. In principle, although FICA and other issues often change the reality, the U.S. federal income tax is set up to be progressive, with tax rates that increase with income. Because this makes federal income taxes somewhat more complex than most of the other taxes we've dealt with, let's focus on them a little more closely.

Understanding the Income Tax

Look back at Einstein's quote at the beginning of this chapter, and then remember that it dates to a time when tax laws were far simpler than they are today. Current tax law (the Internal Revenue Code, also known as Title 26 of the U.S. Code) runs several thousand pages. On top of that come tens of thousands of additional pages of regulations, interpretations, and legal guidance. When you put it all together, it's quite likely that no one—including the members of Congress who write the laws and the IRS employees who enforce it—actually understands all of it.

Nevertheless, the basic idea behind the income tax is fairly simple. You figure out the amount of your income that qualifies as taxable, and your tax is some percentage of that amount. The complexity arises in defining what is taxable, along with the fact that different types of income are subject to different tax rates.

Tax brackets. Let's start with the most basic aspect of taxes—how they are computed *after* you've figured out your taxable income, assuming that it is all "ordinary income" like that you earn from wages or self-employment. The tax you owe is determined by assigning different tax rates to different income ranges, often called *tax brackets*. For example, Table 1 shows the tax brackets for 2011 for a married couple (filing jointly).

Let's see how this works for a married couple that earned $100,000 in taxable income. This amount falls on the third row of Table 1, which shows a tax rate of 25%. However, this rate does not apply to all of their income. Rather, we find their actual tax bill as follows:

- The first row of the table shows that their first $17,000 of income is taxed at 10%, which comes out to $1,700.
- The second row shows that their income from $17,001 to $69,000 is taxed at 15%. The amount of income taxed at this rate is $69,000 − $17,000 = $52,000, so their tax on this income is $52,000 × 15% = $7,800.
- Their remaining income of $100,000 − $69,000 = $31,000 is taxed at the 25% rate, for a tax of $7,750.
- We add the above three tax amounts to find their total income tax: $1,700 + $7,800 + $7,750 = $17,250.

Table 1. 2011 Tax Brackets (married filing jointly)

Tax rate	Taxable income
10%	up to $17,000
15%	$17,001–$69,000
25%	$69,001–$139,350
28%	$139,351–$212,300
33%	$212,301–$379,150
35%	$379,151 and above

Therefore, their overall income tax rate is 17.25% on their $100,000 taxable income. Remember, however, that they will also be paying FICA tax, which can dramatically increase the total rate that they pay.

The key point is that each subsequent level of income is subject to a progressively higher tax rate, which is what makes the overall income tax system progressive. Note, however, that Congress often changes the tax brackets, which changes the degree of progressivity. For example, in 1986, Congress passed and President Reagan signed a tax simplification bill that resulted in only two brackets, one at 15% and the other at 28%. Under President Clinton, a couple of additional tax brackets were inserted, with the top one set at 39.6%. The rates for 2011 were set under the tax law signed by President Bush in 2001. These rates are currently scheduled to expire at the end of 2012.[15] Congress may then either extend them, allow them to expire and return to the pre-2001 rates, or modify them in some other way. We'll discuss some possible modifications shortly.

Gross income, adjusted gross income, taxable income, and tax. The basic idea we've just discussed is simple enough, so now it's time for the complications. The first one is this: Even though I told you the couple earned $100,000 in taxable income, that does *not* mean they actually earned $100,000. Why not? Because taxable income is not the same as total income.

Your actual total income for the year, from all sources combined, is called your *gross income*. You do not pay tax on your full gross income, for two

---

15. Under the tax law passed in 2001, these rates were due to expire at the end of 2010; they were extended for an additional two years through legislation passed in late 2010.

reasons. First, under various special laws, certain portions of some people's income are not subject to taxation, so you are allowed to *adjust* your gross income by subtracting these. The most common adjustment to income is for contributions to individual retirement accounts (IRAs) and other tax-deferred retirement plans; because these contributions are not subject to taxation, you subtract them from your gross income to determine your *adjusted gross income*. Second, most people are eligible for at least a few so-called exemptions and deductions, such as your "personal exemption" or the mortgage tax deduction. These get subtracted from your adjusted gross income to find your taxable income.

With your taxable income finally in hand, you can use tax tables, a list of tax rates like that in Table 1, or tax software to find your tax. But we're not done yet, because you may still be eligible for some *tax credits*, such as for your children or for purchases of solar panels. If so, you subtract the tax credits to find your final tax bill.

Figure 11 summarizes the overall process. Let's take an example. Suppose that our couple actually earned $150,000 from wages, bank interest, and other "ordinary" income sources combined. This $150,000 is their gross income. Now, suppose they contributed a combined $10,000 to tax-deferred retirement plans; we subtract this adjustment to find their adjusted gross income of $140,000. Next, suppose they have two children. For 2011, they were entitled to a personal exemption of $3,700 for each member of their four-person family, for a total of $14,800 in exemptions. This alone reduces their taxable income from $140,000 to $125,200. If they have an additional $25,200 in tax deductions (such as from mortgage interest and charitable donations), it brings their taxable income down to $100,000.

Notice the effect this process has on their "true" income tax rate. The total income tax we found earlier for this couple was $17,250, which was 17.25% of their taxable income. However, it is only 11.5% of their gross income, which sounds a lot less onerous. It could be even lower if they are eligible for any tax credits.

The combination of all these exemptions, deductions, and credits leaves many people owing no income tax at all. As of 2011, about 40% of Americans fell into this category. Of course, most of them still owed FICA taxes, which apply to all wages from the very first dollar, so unless they also qualified for an earned income credit, they still paid federal taxes of some sort.

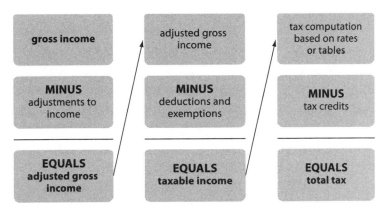

Figure 11. This flowchart shows the basic steps in calculating income tax.

Incidentally, you may notice that adjustments, exemptions, and deductions all have the same basic effect of making your taxable income lower than your gross income. So why the different names? Don't ask me; I'm not Einstein, and I doubt he could have told you either.

Standard versus itemized deductions. When you are figuring out your tax deductions, current tax law gives you a choice. You are automatically entitled to take a *standard deduction*, for which no documentation is required. Or, if it will benefit you more, you can *itemize* your deductions by listing out what you spent on mortgage interest, charitable contributions, and other deductible items. You cannot take both the standard and the itemized deductions; it's one or the other. For example, the standard deduction for married couples in 2011 was $11,600. Therefore, if all your itemized deductions added up to only $11,000, then you were better off claiming the standard deduction.

Tax credits versus tax deductions. Tax credits and tax deductions sound similar but have very different effects on your tax bill. Consider again our couple with $100,000 in taxable income. If they receive a *tax credit* of $1,000, it chops a full $1,000 off their final tax bill. In contrast, a $1,000 *tax deduction* trims $1,000 from their taxable income; because they would have paid a 25% tax on this income, their net savings is only $250. As you can see, tax credits are much more valuable to you than tax deductions (or exemptions or adjustments, which have the same effect).

You may also notice that tax deductions tend to undo some of the progressiveness of the income tax system, because they are more beneficial to higher-income people. For example, suppose our $100,000 couple donates $1,000 to a charity. If we assume that they have enough total deductions so that they itemize (rather than taking the standard deduction), then the deduction they receive for the donation will reduce their tax bill by $250, which means the donation only cost them a net of $750. Now consider a couple that earns only half as much but also donates $1,000 to the same charity. Because this couple is in the 15% tax bracket, the deduction saves only $150 in taxes, making the net cost of the donation $850. This effect is often magnified, because it's likely that the $50,000 couple is taking the standard deduction rather than itemizing. In that case, their $1,000 contribution does not lower their tax bill at all.

The truly mind-boggling aspect of deductions arises when we consider the ones for large expenditures. Consider the deduction for mortgage interest. The average middle-class couple gets little or no benefit from this deduction, because their mortgage interest and other deductions are unlikely to exceed the standard deduction by much, if at all. Now consider a high-earning couple with an expensive house for which the mortgage interest alone is $100,000 per year. Assuming that their income puts them in the 35% tax bracket, the mortgage interest deduction saves them $35,000.[16] In essence, the tax system is subsidizing their expensive home.

Aside from questions of fairness, all these tax breaks have major effects on the overall government budget. The biggest single break arises from the fact that we don't have to pay tax on employer-paid health benefits or self-employed insurance premiums, which saves taxpayers a total of about $300 billion per year. Given that health care is a national priority, perhaps it makes sense for government to make it easier for people to afford. But because of the way deductions work, this break is far more valuable to wealthy executives with "Cadillac" health plans than it is to average people. It's much more difficult to see the sense in that. The same is true for the second-biggest tax break, which goes to contributions to retirement plans. The deduction for

---

16. I'm ignoring numerous other potential complications, such as limitations on the mortgage interest deduction and effects of the AMT. But the basic point still holds.

state and local taxes comes next, followed by the deductions for mortgage interest and charity. In all these cases, the deductions are far more valuable to the wealthy than to the rest. Most amazingly, the government estimates that without these and the many other tax breaks, total tax revenue would be nearly *$1 trillion* higher per year, which by itself would have virtually erased even the record deficits of the past couple years.

Ordinary income versus capital gains. We've already discussed the fact that capital gains from investments are taxed differently from ordinary income. Most dividends are taxed at the same rate as capital gains. There are some good reasons for the special tax treatment of dividends and capital gains, since investments are inherently risky and the government wants to spur more investment. Many countries don't tax capital gains at all. Nevertheless, it's impossible to ignore the fact that the special treatment for capital gains raises some fundamental issues of fairness.

To take a somewhat extreme but illustrative example, consider a single, 20-something rich kid who hasn't done much productive with his life, but who is fortunate enough to have income from an inheritance that provides him with dividends and capital gains totaling $150,000, the same as the gross income for our married couple above. Let's say that exemptions and deductions get his taxable income down to $100,000, also matching our married couple's. At 2011 rates, his tax on this income would be $9,825 (because the first $34,500 of the dividends/capital gains are untaxed, and the rest is taxed at 15%). Moreover, he would not owe any FICA tax, because the income was not "earned." Therefore, his total federal tax bill would be $9,825, or 6.55% of his $150,000 gross income.

Now let's go back to our married couple with two children and assume that their $150,000 gross income was entirely from wages, evenly divided between the two of them so that both of their individual incomes fall below the threshold at which FICA cuts off. In that case, they will pay FICA tax of 7.65% on the entire $150,000, which comes out to $11,475. Adding this to the $17,250 of income tax that we found for them earlier, their total federal tax bill is $28,725, or 19.15% of their gross income.

Whatever you may think about the general wisdom of taxing capital gains, it's difficult to see any fairness in the idea that a person living off an inheritance should pay only a third as much in tax as a hardworking couple with the same total income. Indeed, as we saw in our chapter-opening ranking task, the couple is also paying a higher tax rate than a billionaire investor. I know that economists may argue the point with me, but I believe that society as a whole would benefit much more from having a tax system that is perceived as fair than it does from any calculations about the economic merits of taxing capital gains differently from income earned through work.

How Should We Change Our Tax System?

When you look at the complexities of the current system and add in the issues of fairness, there's no doubt that there has to be a better way. The more difficult question is deciding which of many better ways we should go.

The good news is that, as this book goes to press (late summer 2011), a number of reform proposals are on the table. Several high-profile commissions have proposed changes designed to simplify taxes and help with our national deficit/debt situation, and congressional Republicans and President Obama have offered their own plans. The bad news is that, based on the recent history of major reform proposals, politicians are unlikely to approve any of these plans. We can always hope that this time will be different, and the debt ceiling deal passed in August 2011 calls for such reform, but I'll have to see it to believe it.

Current proposals aside, I believe that the quantitative reasoning we've applied to understanding our current tax system can also help point the way to a better system. Although I'm now delving into the realm of opinion rather than mathematical truth, I'll list out five basic principles that I personally believe our tax system should follow.

First, I think we should couple tax reform with efforts to balance the federal budget, so that we will not continue to pass costs on to future generations. Better yet, we should seek a small surplus, so that we can gradually pay down the existing federal debt.

Second, I believe the tax system should be much simpler. With rare exceptions, it should be possible to fill out your income taxes on a postcard, and it certainly shouldn't be necessary for the average person to need help from an accountant.

Third, I believe it's important for people to perceive that the system is fair. To that end, I think we should strive for a system in which two people who earn the same amount of money pay the same amount of tax, regardless of the source of their income or of the personal decisions they make about how to spend their money. I would therefore eliminate the distinction between "ordinary" income and capital gains, since we've already seen that this system leads to bizarre consequences such as billionaires and trust fund kids paying lower tax rates than working people. (However, if long-term capital gains are to be taxed as ordinary income, it probably makes sense to index them for inflation.) I would also eliminate *all* tax deductions, which would have the side benefit of allowing us to lower overall tax rates. For those who worry that eliminating the mortgage tax deduction would hurt the housing market, look back at Chapter 1 and I think you'll see that it's just as easy to argue that this deduction helped cause many of our recent economic problems. For non-profit supporters who worry about elimination of the deduction for charitable donations, keep this fact in mind: On average, poorer people tend to give a higher percentage of income to charity than wealthier people, yet they get no tax benefit from this giving (since nearly all of them take only the standard deduction). There's no way to know for sure, but I'm hopeful that the lower tax rates we'd have by eliminating deductions would put people in a mood to be more generous overall.

Fourth, the tax system should be more honest, eliminating accounting games that obscure the true rates that people pay. This principle applies in particular to the current distinction between "income" taxes and FICA taxes, which as we've seen are both taxes on income. The common argument for keeping the two separate is that the latter is supposedly a kind of national retirement plan. However, as we'll discuss in the next chapter, the government mixes income and FICA taxes together when it spends our tax money, and the amounts that current retirees are drawing from the system are not clearly representative of the amounts they paid in. The relationship between contributions and benefits is likely to be even weaker in the future, largely because the government is no longer collecting enough in FICA taxes to cover the benefits that it pays out. With these facts in mind, I believe we should simply decide what we're going to spend as a nation and collect what we need to cover that spending in taxes, without making artificial distinctions between tax types that can cause middle-class people to end up paying higher overall income tax rates than much wealthier people.

Finally, I don't believe that there's any getting around the fact that it's much more difficult for a poor person to give up a dollar of income than it is for a richer person. Richer people also tend to get more benefit from having a well-functioning government; after all, it's much more difficult to live a good life in a more corrupt or less democratic country, no matter how much money you have. For those reasons, I believe that we should retain a progressive tax system. Nevertheless, no one should have to pay so much that they lose incentive to work, so we should try to keep the top rate as low as possible while still meeting all our budget obligations. It's worth noting again that elimination of deductions could allow us to lower tax rates significantly without a loss of government revenue.

These proposals focus exclusively on income taxes. Some people argue that we should replace some or all of our federal income taxes with a consumption tax, such as a national VAT or the consumption-based "FairTax" that has been proposed in Congress. I see great merit in the principle behind this idea, because it tends to encourage savings over consumption, and can make it more difficult to game the system or cheat to avoid taxes. However, a consumption tax also has downsides, including the burden it places on small businesses and its tendency to be regressive unless complexities are added to prevent the poor from paying more than the rich. Although I'm sure you've figured out by now that I tend to be highly opinionated, I find myself unable to make a clear decision as to whether the benefits of a consumption tax would outweigh its drawbacks. For that reason, I take no position on it for now, but I think it at least deserves further consideration as a possible solution to the budget crisis that we find ourselves in.

Math for Life

Taxes are a critical issue for almost everyone, yet many people are confused about how taxes actually work. To take one simple example: Large numbers of people would look at the tax rates in Table 1 and conclude that a couple earning $100,000 per year is paying 25% of that income in taxes. In other words, they don't realize that *taxable* income is very different from actual income, that the 25% applies only to a portion even of taxable income, and that it represents only part of actual income tax, since it does not include FICA taxes.

This lack of understanding of the basic principles of our tax system leads to many other fundamental misconceptions and probably explains a lot of the overheated debate about taxes. On the liberal side, it's rare to hear it acknowledged that taxes place a real burden on people's finances, or that high tax rates can be a disincentive to work or an incentive to game the system. On the conservative side, it's common to hear complaints that 40% of Americans pay no "federal income tax," ignoring the facts that most of these people still pay FICA taxes and that they don't earn very much money in the first place.

Of course, given the complexities of the current system, the lack of public understanding is not too surprising, and the many instances in which the system is obviously unfair probably contribute greatly to public anger over taxes. After all, it's bad enough that you have to pay a lot of money in taxes, but it's far worse if you think that you're paying more because others are not paying their fair share.

All that said, I think it's also important to take a deep breath and look at how we fare overall. No one likes paying taxes, and there's no doubt that there's lots of wasteful government spending and many government programs that could or should be eliminated entirely. Nevertheless, this is still a pretty great country. I would argue that there's never been another time or place in history in which human beings have been able to live quite as comfortably and freely as we can in the United States today. So the next time you find yourself angry at a tax bill, step back and ask yourself: Even if they'd waive all your taxes, would you be willing to relocate to Libya, or Afghanistan, or Congo? Do you really believe that you'd still be able to live just as safely and comfortably, and earn just as much money, in these and other places? I'm reminded of a famous quote from Supreme Court Justice Oliver Wendell Holmes: "I like paying taxes. With them I buy civilization."

I believe that if we keep this sense of balance in mind, and use our skills of quantitative reasoning, then we'll all be able to tone down the rhetoric and find a way to make a tax system that is better and fairer, while at the same time ensuring that government spends our tax money more wisely. In this sense, understanding taxes is a classic example of "math for life," and one that we should all seek to do better at.

6

The U.S. Deficit and Debt

There are only two kinds of taxes, the ones we pay now and the ones we pay later or defer to our children and grandchildren, with interest.
— **David Walker, in *Comeback America***

We Americans, all of us, have a problem greater than . . . anything else we face, including the dreadful threat of terror. The problem is our national debt.
— **John Danforth, former U.S. Senator (Republican–Missouri)**

Question: Consider a family of four earning about $100,000 per year. Based on U.S. averages, the tax cuts enacted in 2001 reduced this family's tax bill by about $1,500 per year over the next 10 years. How did the family's net lifetime tax liability change during the 10-year period? (Assume U.S. averages in choosing your answer.)

Answer choices:
a. Their lifetime net tax liability fell by $1,500.
b. Their lifetime net tax liability fell by $15,000.
c. Their lifetime net tax liability fell by $15,000, plus interest.
d. Their lifetime net tax liability increased by about $105,000, plus interest.

If you answer this question based solely on the numbers I've given, you'll conclude that the answer is either B or C. After all, if a family's tax bill goes down by $1,500 per year, then over 10 years they'll end up with $15,000 still in their pockets that otherwise would have gone to taxes. If they were able to save or invest that money, then they'd have earned interest on it too. But this is a trick question, because I asked what happened to their *lifetime* tax liability, and that requires thinking about some numbers that I didn't give you.

The key idea is found in the chapter-opening quote from David Walker. In 2001, the federal debt stood at about $6 trillion. A decade later, the debt had risen to almost $15 trillion. If we divide the $9 trillion increase in the debt by the U.S. population of about 310 million, it works out to almost $30,000 per person, or $120,000 for a family of four. This money will eventually have to be paid back with money from taxpayers. Therefore, while the family did indeed save $15,000 during the 10-year period, their future tax liability went up by $120,000, making a net increase in their liability of $105,000. We also have to pay interest on the national debt, so the correct answer is D.

There's great political debate over who should be blamed for the huge increase in the debt. Democrats tend to put much of the blame on the tax cuts enacted in 2001, while Republicans tend to put the blame on too much government spending. But no matter where you place the blame, there's no denying the insanity of the result.

Let's use an analogy. Suppose you want to buy a new car, and to do it you decide to take out a $15,000 loan. You'll obviously expect to pay back the $15,000 with interest. But imagine that, one day, you look at your loan statement and find out that your loan balance isn't $15,000 plus interest, but instead is $120,000 plus interest. Thinking it's a simple error, you call the bank to get it fixed, and you learn that the fine print on your loan made you and other customers liable if the bank got into any trouble for any reason, and some poor management decisions have put the bank deep in a hole. You'd be livid, because you never would have agreed to the loan if you'd understood the meaning of that fine print. Yet, as a country, we've accepted this same fine print in allowing the government to spend more than it collects in taxes, causing our net lifetime tax liabilities to spiral out of control.

Our future as a nation depends on our ability to regain control of national finances. Much as we did in Chapter 4 in discussing personal finances, we'll begin by looking at the general budget process.

National Budgeting

In principle, the national budget process is not much different from your personal budget process. We simply add up all the government's income and subtract all its expenditures to find its net cash flow, which is called a *deficit* if it is negative and a *surplus* if it is positive. The major difference between the government's budgeting and yours is in the number of people required to agree. Instead of just you (and perhaps your spouse or family members), the government budget has to be agreed upon by majorities of the 435 members of the House and of the 100 members of the Senate, and then signed into law by the president. Since these people face elections, they come under great pressure to please their constituents, and much of our current predicament can be traced to politicians who aim to please so much that they ignore the consequences of their actions. As John Danforth (also quoted at the beginning of this chapter) put it, "Benefits are popular. Paying for benefits is extremely unpopular."

Deficit and Debt

The terms *deficit* and *debt* are easy to confuse, but there's an important difference between them. The deficit is the shortfall in the government's cash flow in any single year. The debt is the total amount of money that the government is obligated to repay. In other words, each year's deficit adds more to the total debt.

Deficits and debt have been with us for a long time. The left graph in Figure 12 shows annual deficits or surpluses from 1971 through 2010; be sure to notice the break in the vertical scale, which otherwise masks a huge increase in the deficit for the last two years shown. With the notable exceptions of the years 1998 to 2001, the government ran deficits every year. As a result, the debt, shown in the graph to the right, grew every year, except for the almost-plateau during the short surplus period. (Even then, the debt rose a bit, for reasons we'll discuss shortly.) The budgetary crisis we now face comes from the combination of the long-term accumulation of debt and the record deficits of the past decade, which show no signs of abating.

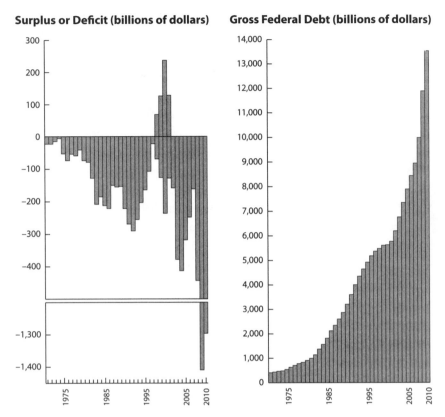

Figure 12. Annual deficits (left) and total debt (right) for the U.S. federal government, 1971–2010. Note that all values are in *billions* of dollars; the numbers are *not*
adjusted for inflation. Source: U.S. Office of Management and Budget.

It's worth noting that the current concerns followed a period of great opti-
mism for the federal budget. In early 2001, economists predicted that the government would achieve a combined *surplus* of $5.6 trillion over the next ten
years, which would have been nearly enough to retire the debt by now. Instead,
as we've already discussed, the debt ended up rising by some $9 trillion over
that same period. In other words, the difference between the projection and
the actual turned out to be almost $15 trillion. You might wonder how the
economists of early 2001 could have made such a huge error. The answer is
that their projections were based on mathematical models of the economy,

and the inputs to those models included several key assumptions that ended up not coming true. For example, they assumed that tax rates would stay where they were at the time, but instead taxes were cut dramatically just a few months later. They also assumed the economy would remain strong, rather than being hurt by the terrorist attacks of that year, and of course they did not account for the costs of the wars that followed those attacks or for the effects of the recent recession. There's no way to know whether we'd really be living in a debt-free nation if those events had not occurred, but this provides an important lesson in mathematical modeling: The predictions of models are only as good as the assumptions that go into them. Or, as attributed to Niels Bohr, Yogi Berra, and others: "It's tough to make predictions, especially about the future."

If your personal budget looked as dire as the national budget, you'd be forced to make some tough decisions, and that would require understanding both your sources of income and your spending patterns. The same is true for the government budget, so our next step is to understand the inflows and outflows of taxpayer money.

Following the Money

The pie charts in Figure 13 show the basic sources of the government's income and the makeup of its expenditures. On the income side, notice that more than 80% comes from the combination of individual income taxes and the FICA taxes collected primarily for Social Security and Medicare, and half of the rest comes from corporate income taxes. This makes it pretty easy to see the options for raising more revenue. While increases in things like gasoline taxes and "sin" taxes (which count as "excise taxes" in the pie chart) could help a bit, the only ways to raise significant amounts of additional revenue are through increases in taxes on either individuals or corporations. Many economists believe the corporate income tax should be eliminated, both because it's generally passed on to consumers anyway and because it harms the global competitiveness of U.S.-based companies. In that case, any significant new revenue would ultimately have to come from either increases in individual income taxes or new individual taxes, such as a national sales tax or VAT.

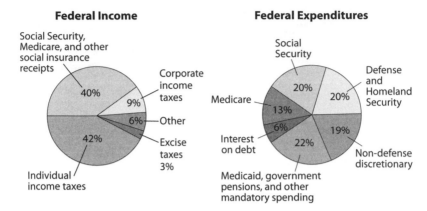

Figure 13. Approximate makeup of federal income and expenditures for fiscal year 2010. Source: U.S. Office of Management and Budget.

Of course, we wouldn't need to raise more revenue if we could cut spending enough, but the spending pie chart shows why this is very difficult to do. Broadly speaking, government spending falls into two major categories:

- "Mandatory" expenses are those that are paid automatically *unless* Congress acts to change them. They include Social Security, Medicare, Medicaid, and government pension programs, since all these programs pay out money based on promises made by the government in the past. Interest on the debt is also a mandatory expense, since it represents a promised payment by the government to those who hold the debt. Notice that all but the two right-most wedges in the pie chart are mandatory expenses.

- "Discretionary" expenses are the ones that Congress must vote on each year. In the spending pie chart, discretionary expenses are subdivided into those that affect national defense and security, and all the rest ("non-defense discretionary"), which includes education, roads and transportation, agriculture, food and drug safety, consumer protection, housing, the space program, energy development, scientific research, international aid, and virtually every other government program you've ever heard of (and many that you haven't) that we haven't already mentioned.

You can probably see the problem. The pie chart shows mandatory expenses making up more than 60% of current government spending, and this percentage is expected to rise substantially in the future, primarily because of an increasing population of older Americans (which means higher costs for Social Security and Medicare) and rising payments for interest on the national debt. Projections vary, but let's suppose that mandatory expenses are on track to grow 50% in the next decade or two. Then, if we wanted to hold total spending constant, the share of federal spending going to these programs would have to rise from about 60% to about 90%—which means the portion going to discretionary spending would have to drop to 10%. Notice that even if we cut defense spending in half, that would leave *nothing at all* for education, roads, scientific research, or any other non-defense discretionary programs.

Just to hammer home how badly the current budget is broken, consider the "do nothing" option. If we don't either raise revenue or cut spending, then the debt continues to grow, and a growing debt means rising payments for interest on the debt. In fact, our annual interest payments are already expected to more than double during this decade. What does doubling of the interest payments mean? It would take them to more than $800 billion per year—considerably more than the projected total for *all* non-defense discretionary spending. Most of the interest goes to wealthy investors and foreign governments, which means we're effectively giving these groups money that we otherwise could have spent on things like the education of our kids—the very education they'll need to make smarter budgetary decisions than their parents did.

Facing Budget Reality

Something must be done, but the political left and right have been equally blind to this reality. People on the left usually advocate changes such as cutting defense spending and increasing taxes. But even drastic cuts in defense wouldn't be enough unless we also raised taxes to untenable levels. People on the right want to focus on cutting spending, but with some prominent exceptions, they usually oppose cuts to defense, and polls show that even many Tea

Party supporters oppose cuts to Social Security and Medicare. That leaves only non-defense discretionary spending, and as we've seen, even cutting that to zero wouldn't solve the problem.

The clear reality is that we can solve the budget problem only if we put everything on the table, including tax increases and major cuts to mandatory programs like Social Security and Medicare. This will not be popular with anyone, but that's as it should be. After all, if you found your personal budget in circumstances anywhere near as bad as the government budget's, you'd probably recognize that there was no way out without some real pain. So it's time to face reality, and face the pain. If we don't, even greater pain will fall on our children and grandchildren.

Debt, Trust Funds, and Social Security

Earlier, I briefly noted that the debt rose even during the period from 1998 to 2001 when the government ran annual surpluses rather than deficits. Moreover, if you look closely at the deficit and debt graphs, you'll find that the debt rises by *more* than the amount of the deficit each year. If you're like I was when I first encountered these numbers, you may be a bit mystified. To remove the mystery, we must look a little more closely at how the government finances its debt.

Two Types of Debt: Publicly Held and Gross

If your personal budget is in the red, you must cover the shortfall either by withdrawing from savings or by borrowing money. The federal government does both. It withdraws money from its "savings," and it borrows money from people and institutions willing to lend to it.

Let's consider borrowing first. The government borrows money by selling debt to the public in the form of Treasury bills, notes, and bonds, jointly called Treasury "issues" (because they are issued by the Department of the Treasury). If you buy one of these Treasury issues, you are lending the government money that it promises to pay back with interest. Because Treasury issues are considered to be very safe investments,[17] the government has never yet had trouble finding people, institutions, or foreign governments willing to buy them. By the end of 2010, the government had borrowed a total of about $9 trillion through the sale of this debt, which is usually called the *publicly held debt.*

The government's "savings" consist of special accounts, called *trust funds,* which are supposed to help the government meet its future obligations to mandatory spending programs. The biggest trust fund by far is for Social Security. Over the past few decades, FICA taxes have brought in much more money for Social Security than has been paid out in benefits. Legally, the government has been required to invest this excess money in the Social Security trust fund, so that it would be there when it was needed for future retirees. But there's a catch: Before the government borrows from the public to finance a deficit, it first tries to cover the deficit by borrowing from its own trust funds.

In fact, the government has to date borrowed every penny it has ever deposited into the Social Security trust fund, and the same is true for other trust funds. In other words, there is *no actual money* in any of the trust funds; instead, they are filled with the equivalent of a stack of IOUs (more technically, with Treasury bills) representing government promises to return the money it has borrowed, with interest. At the end of 2010, the government's debt to its own trust funds was about $4.5 trillion. Adding this to the $9 trillion publicly held debt gives a *gross debt* of about $13.5 trillion for 2010, which is what Figure 12 shows.

The trust funds explain the strange fact that the debt can rise even in surplus years. Let's take the last surplus year, 2001, as an example. The surplus that year was $128 billion, meaning that the government really did collect $128 billion more than it spent. However, based on the excess the government collected in Social Security taxes (compared to the benefits it paid) and

---

17. This safety is due partly to the fact that the government can in principle always pay its debt, since it has the power to print money. There is a legal complication, however: The government's total debt is required to remain below a "debt ceiling," which means that every time the debt approaches this ceiling, Congress must vote to raise the ceiling. As this book goes to press, the debt recently reached the ceiling, and Congress passed an increase just barely in time to avert a possible default that might have changed the perceived safety of Treasury issues.

the interest it owed the trust fund for its past borrowing, the government was legally required to deposit $161 billion into the Social Security trust fund that year. As you've probably guessed, the government met this obligation not by depositing real money, but rather by adding that much in IOUs (in the form of Treasury bills) to all the other IOUs in the fund. Now, notice that this $161 billion in IOUs represents $33 billion *more* than the surplus for the year. In other words, if you include this obligation, the government actually recorded a $33 billion deficit for the year.[18] The government borrowed similarly from other trust funds, and the net result was that the gross debt rose by $141 billion, despite the surplus recorded for the year.

The Hard Truth of Social Security

With all this talk of government giving IOUs to itself, you may be starting to think that something's fishy with Social Security. You'd be right, but you haven't seen the half of it yet. Social Security is usually presented to the public as a retirement plan, in which you pay in through your FICA taxes and later collect based on what you put in. The Social Security Administration even provides handy little reports of how much you've put in and the payments you'll therefore be entitled to when you retire. This tends to make current retirees think that they're just getting back their own money each month, and it also explains why many people think FICA taxes are somehow different from other income taxes. But while there's some linkage between past payments and future benefits, the linkage is quite weak. Today, most retirees collect far more in benefits than they ever paid in, even when you account for inflation, in part because longer lifetimes mean many more years spent in retirement. Moreover, Social Security is not just for retirees; it's also a disability program, and the difference between past payments and future benefits is even greater for people in this category.

---

18. You may wonder how the government can claim it had a surplus when its obligations to the Social Security trust fund mean it actually ran a $33 billion deficit. The answer is that current law allows the government to treat Social Security as "off budget," which essentially means the government doesn't have to count it when it reports its annual deficit or surplus.

The truth is that while Social Security acts somewhat like a national retirement and disability plan—it currently provides monthly payments to more than 50 million retired and disabled workers—it is not at all like an individual retirement plan, in which you deposit money into an account while you work and withdraw it later when you retire. This difference between Social Security and personal retirement plans isn't necessarily a bad thing, and it might actually work quite well if the Social Security trust fund held real money. But as we've discussed, it's actually filled with IOUs.

We can use an analogy to see just how bad this situation is. Imagine that, at a young age, you decide to set up a retirement savings plan that will allow you to retire comfortably at age 65. Based on your best guesses about future interest rates, you calculate that you can achieve your retirement goal by making monthly deposits of $250 into your retirement plan. So you start the plan by making your first $250 deposit.

However, the very next day, you decide you want a new TV and find yourself $250 short of what you need. You therefore decide to "borrow" back the $250 you just deposited into your retirement plan. Because you don't want to fall behind on your retirement savings, you write yourself an IOU promising to put the $250 back. Moreover, recognizing that you would have earned interest on the $250 if you'd left it in there, at the end of the month you write yourself an additional IOU to replace this lost interest.

Month after month and year after year, you continue in the same way, always diligently depositing your $250, but then withdrawing it so you can spend it on something else, and replacing it with IOUs for the withdrawn money and the lost interest. When you finally reach age 65, your retirement plan will contain IOUs that say you owe yourself enough money to retire on—but you'll obviously find it very difficult to live off them.

No rational person would ever treat a retirement plan in the way we've just described, yet this is essentially the way the Social Security trust fund works: The government has been diligently depositing the excess money collected through FICA into the Social Security trust fund, then immediately withdrawing it for other purposes while replacing it with Treasury bills that are nothing more than IOUs. It was easy to ignore this insanity as long as the IOUs kept piling up without anyone trying to collect on them. But that has now changed. High unemployment has lowered FICA tax collections, so that in 2010, for the first time, Social Security payments exceeded Social Security tax collections. The difference was small enough to barely register in the federal deficit, but this tolerable situation won't last. Without changes, Social Security's payments are expected to exceed its tax collections by ever-larger amounts in the coming decades.

To see the problem vividly, consider the year 2036, which is approximately when the government's "intermediate" projections (meaning those that are neither especially optimistic nor especially pessimistic) say the Social Security trust fund will finally run dry. That year, projected Social Security payments will be about $600 billion more than collections from Social Security taxes,[19] which means the government will be redeeming its last $600 billion in IOUs from the Social Security trust fund. But since the government owes this money to itself, it will have to find some other source for this $600 billion. Generally speaking, the government could find this money through some combination of the following three options: (1) it could cut spending on discretionary programs; (2) it could borrow the money from the public by selling more debt (in the form of Treasury bills, notes, and bonds); or (3) it could raise other taxes.

Unfortunately, none of the options are viable. The $600 billion is more than the total amount of *all* non-defense discretionary spending, so the government would have to eliminate all these programs and substantially cut the military to save this much money. Borrowing an additional $600 billion might be viable if it were a one-time event, but these IOU redemptions will be an ongoing trend; I don't believe there's any chance that U.S. Treasuries would still be considered sound investments if we kept borrowing so much year after year. An increase in other (non-FICA) taxes won't work either; if you go through the analysis, you'll find that the government would have to raise individual income tax revenues so dramatically to collect an extra $600 billion that the tax increases might well wreck the economy. We might try to get $200 billion from each of the three sources, but I'm afraid that wouldn't work either, because the cumulative effects over many years would be too devastating to both the government budget and the economy.

---

19. This and other numbers cited in future projections are given in current (2011) dollars, so you don't need to think about how they'll be affected by future inflation.

That's not even the worst of it. Medicare is expected to face a similar crisis, and many economists believe that rapidly rising health costs will make that crisis much more severe than the one for Social Security. Moreover, I'm skeptical that the trust fund will last even to 2036 (which is still long before today's young workers will retire). Why am I skeptical? Because of life expectancies. When predicting future Social Security payments, the government has to make assumptions about future life expectancies; after all, if people live longer, then more people will still be alive collecting benefits in the future. To take an example, the current projections assume that American women will not reach a life expectancy of eighty-two until about 2030—but that's *already* the life expectancy of women in France. In fact, during the twentieth century, U.S. life expectancies rose an average of about three years per decade. If that trend continues, life expectancy for women will reach eighty-six by about 2030, which throws the current Social Security (and Medicare) projections almost completely out the window.

The bottom line is that the Social Security trust fund is a myth. People can (and do) argue about whether it is fair or right that Social Security money has been used for other purposes, but this won't change the reality that the trust fund contains no money. As a result, the system is headed for dramatic failure, and we either act to fix it now, or suffer the consequences later.

Math for Life

Look again at the two quotations that open this chapter. David Walker's quote is undeniable; when we accumulate debt, we are only passing the taxes to pay for it on to future generations. John Danforth's quote is a little more debatable, since we also face other major problems (we'll discuss some in the coming chapters), but there's no denying that our national debt ranks as one of the worst.

What may set this problem apart from most others is that the outlines of the solution are fairly obvious. We need to make sure we don't keep spending more than we take in. I don't believe that can be done either by raising taxes alone or by cutting spending alone, so it will have to be accomplished by some combination of both. I've already addressed (in the last chapter) how I think we could improve our tax system, and increasing revenues is just a matter of setting tax rates at the right levels. On the spending side, like a cash-strapped family, we need to put our priorities in order and decide what we can keep and what can go. Personally, I'd start by getting rid of the distinction between mandatory and discretionary expenses, turning programs like Social Security and Medicare into safety nets for those who truly need help, rather than entitlements for everyone. You may have a different opinion, but if we all faced reality, I'm sure we could come to a compromise about what to cut.

That brings us back to the question of why we haven't yet faced reality. In keeping with the theme I introduced in Chapter 1, I think the answer is clear: As a nation, we're being very "bad at math." We can't simply blame our politicians, since they do what their constituents ask them to do. We have to put the blame on ourselves, both for failing to accept the severity of our national budget predicament, and for being unwilling to work out the details of the only possible solution. And as usual, I place much of the blame for our blinders on our educational system. These budget problems have been building for decades, yet they're still rarely taught anywhere outside of a college economics department.

In this case, however, we don't have time to wait to improve our educational system. We need to act now. Please, tell your friends and neighbors, and tell your politicians: We know the solution is going to hurt, but we can't hand these problems over to our kids.

7

Energy Math

We've embarked on the beginning of the last days of the age of oil. . . .
Embrace the future and recognize the growing demand for a wide range of
fuels or ignore reality and slowly—but surely—be left behind.
— **Mike Bowlin, chairman of ARCO (now part of BP), 1999**

One reason we use energy so lavishly today is that the price of energy
does not include all of the social costs of producing it.
— **President Richard Nixon, 1971**

Question: Suppose that we had the ability to generate power through nuclear *fusion* (which is different from the fission used in current nuclear power plants), using hydrogen extracted from ordinary water as the fuel. If you had a portable fusion power plant and hooked it up to the faucet of your kitchen sink, how much power could you generate from the hydrogen in the water flowing through it?

Answer choices:

a. Enough to provide for all the electricity, heat, and air conditioning you use in your house

b. Enough to provide for the energy needs of everyone on your block

c. Enough to provide for the energy needs of approximately 500 homes

d. Enough to provide for the energy needs of approximately 5,000 homes

e. Enough to provide for all the energy needs of the entire United States

This question is what is known as an "order of magnitude" question, meaning that instead of looking for an exact answer, we are looking only for a general sense of the size of the answer. That is, we're not concerned with knowing the answer to within 10% or 20%, only with knowing it within a factor of 10 or so. At first, knowing something only within a factor of 10 might seem like knowing very little, but it can often be quite meaningful. For example, a business will operate very differently if it estimates its potential customer base at 1,000 people than if it estimates it at 100 or 10,000. In the same way, each answer choice to our fusion question would give a very different sense of the ways in which fusion might be useful.

So how can you determine the answer? One way might be to guess. If you're like most students I've worked with, you'd then choose randomly among choices A through D, since E sounds like an implausible throwaway. But you probably know that I don't really want you to guess; instead, let's think about how we could actually figure it out. There are only two steps.

First, we need to know how much water flows through your kitchen faucet. The easiest way to learn this is to turn it on, place a pitcher under it, and see how much water you can collect in a fixed amount of time. You'll find that a typical kitchen faucet pours out about three quarts of water per minute, or just over one and a half ounces per second. The second step is to calculate the amount of energy that can be generated by fusing the hydrogen in this water. This step takes a little more work, since it requires data about the amount of hydrogen in each ounce of water and the amount of energy released by fusing that hydrogen. You can find the necessary data easily on the Web, but I'll just tell you the answer: If you could fuse all the hydrogen in the water flowing from your kitchen faucet, you would generate about three *terawatts* of continuous power.

What are terawatts? A terawatt is a trillion watts, which means that continuous power of one terawatt is enough power to keep 10 billion 100-watt lightbulbs, or the equivalent thereof, turned on for as long as the power is turned on. So three terawatts is enough power to light the equivalent of 30 billion 100-watt lightbulbs . . . which turns out to be roughly equal to the total amount of continuous power used by the entire United States. In other words, the "throwaway" answer wasn't a throwaway at all.

Think about what we've just found out. If we had the technological capability for fusion power, and if you were willing to allow us to borrow your kitchen sink, then we could stop drilling for oil, we could stop digging for coal, we could dismantle all the dams on our rivers, we could take down all the wind turbines, and we could even turn off all the currently operating nuclear power plants. All we'd need is the power coming from fusion at your kitchen sink.[20] We'd use that power to generate electricity, which would in turn power all our cars, homes, and industry.

The potential of nuclear fusion is truly mind-boggling, an example of what is sometimes called a game-changing or disruptive technology—one that would fundamentally change the way we approach energy, both economically and politically. So why aren't we building nuclear fusion power plants? Because we don't yet know how. Whether we can figure it out soon enough to make a difference to our current energy problems is a topic of great debate, and one that we'll return to later in this chapter. But I hope I've at least opened your mind a bit to what may be possible if we work hard enough. As Mike Bowlin says in the quote that opens this chapter, a key to solving our problems lies in finding ways to "embrace the future."

Understanding Energy

The first step in understanding our energy problems is to understand energy itself. We use energy in many different forms. Our bodies use energy from food to keep our hearts beating, to contract the muscles we use in breathing and walking, and to generate the electrical and chemical impulses that allow us to think. Cars use energy from gasoline to move pistons in their engines, which turn the wheels so the cars can go. Lightbulbs use energy coming in the form of electricity to generate light, and solar panels collect the energy of sunlight.

---

20. Fusion turns out to be somewhat easier if, instead of using ordinary hydrogen, we use an isotope of hydrogen called deuterium. About 1 in 50,000 hydrogen atoms is deuterium, so to use deuterium we'd need a water flow equivalent to about 50,000 kitchen sinks—which is still only about the flow rate of a small creek. (Fusion is even easier with helium-3, but we'd have to get that from the Moon—which may not be as ridiculous a suggestion as it might at first sound.)

We can simplify our discussion of energy by categorizing it into three basic types: energy of motion, energy of light, and stored energy. Energy of motion is the energy possessed by anything that is moving, whether it is a car driving down the highway, a beating heart, or the molecules that are constantly moving in the air around you. Energy of light is contained in light itself and varies with the type (or wavelength) of light. For example, visible light has just enough energy to activate the cells in your eye that allow you to see, while ultraviolet light has more energy and can therefore damage skin and cause cancer. Stored energy, more formally called "potential energy," includes the chemical energy stored in food, the electrical energy carried by power lines, and the energy stored in mass itself. This last form is quantified in Einstein's famous equation $E = mc^2$, in which E stands for energy, m stands for mass, and c stands for the speed of light. If you put in the measured value of the speed of light (186,000 miles per second, or 300,000 kilometers per second), then you can use this equation to calculate how much energy is stored in any particular amount of mass.

Despite its different forms, energy has an underlying simplicity that is expressed by the *law of conservation of energy*: Energy can neither be created nor destroyed; it can only be converted from one form to another. When your body uses the energy of food to make your heart beat, it is converting the stored energy of the food into the motion energy of your heart. When you turn on a lightbulb, the stored energy of electricity is used to make the bulb light up. Nuclear energy comes from converting energy that is stored in mass into electricity.

These simple ideas are already enough to allow you to understand the basics of our energy economy, which we can summarize as follows: We need energy to run all the machines, appliances, and lighting that we use in our modern society. This energy cannot be created magically, so it must come from somewhere, which means we must find naturally existing energy sources.

Measuring energy. If we're going to discuss our energy problem quantitatively, we first need to discuss how we measure energy. Just as we can measure lengths in inches, feet, or meters, we can use a variety of different units to measure energy. For Americans, the most familiar energy unit is the *food* *calorie*; the average adult needs about 2,500 of these each day. In the rest of the world (and in science), where the metric system is preferred, the standard unit of energy is the *joule*. One food calorie is equivalent to 4,184 joules, or 4.184 kilojoules. If you buy a candy bar or can of soda in Mexico or Europe, the label will show you its energy content in kilojoules rather than calories.

The difference between energy and power. Energy and power are not quite the same thing, and understanding the difference between them is important to being able to converse about energy issues. The difference is this: Power is the *rate* at which energy is used. Those of you who work out at a gym can probably find electronic readouts on treadmills or bicycles that will make the difference clear. For example, suppose you are riding a bike and the readout says that your pace represents 200 calories per hour. This is a measure of *power*, because it tells you the rate at which you are converting energy that was previously stored in your body into energy that turns the pedals; note that the units are energy (calories) divided by time (hours). A second read-out may tell you the total energy you've converted so far. If you've been riding for a half hour, then you have converted 100 calories of energy so far; if you've been riding for two hours, then you've converted 400 calories of energy.

A similar idea applies to the energy you buy from the electric company. Consider a 100-watt lightbulb. It obviously requires more energy to keep this lightbulb on for two hours than for one hour. But it's a 100-watt lightbulb at all times, which tells us that watts must be a unit of power. In fact, a watt is defined as 1 joule per second, which means that a 100-watt lightbulb requires 100 joules of energy if you leave it on for one second, 200 joules if you leave it on for two seconds, and so on. In principle, then, the electric company could send you a monthly bill telling you how many joules of energy you used in your home, and charging you some price per joule. In practice, electric bills more commonly state your energy usage in a different unit: *kilowatt-hours*, often abbreviated "kwh." Because "kilo" means 1,000, a kilowatt is 1,000 watts, and a kilowatt-hour is an hour's worth of energy using 1,000 watts of power. Another example should make this clear.

Suppose that, after sunset, the lights you have turned on in your house are using a total of 1,000 watts, or 1 kilowatt, of power. If you leave the lights on for four hours, they require a total of 4 kilowatt-hours of energy; if you do this every night, it adds up to about 120 kilowatt-hours per month. Now check your electric bill, which will show the price you pay per kilowatt-hour of energy. If the price is a fairly typical 15¢ per kilowatt-hour, then the cost of operating the lights is 120 × 15¢ = $18 per month, or about $216 per year. You can now see the advantage of switching to lightbulbs that use less energy, such as compact fluorescents (CFLs) or light-emitting diodes (LEDs). These lights typically require only about one-quarter as much power to produce the same amount of light as standard lightbulbs. Therefore, if you replace 1,000 watts' worth of ordinary bulbs with CFLs or LEDs, you'll reduce your annual lighting bill about 75%, from $216 to a little over $50. Although the new bulbs are more expensive than ordinary bulbs, they'll quickly pay for themselves in energy savings.

BTUs and beyond. The list of acceptable energy units doesn't end with calories, joules, and kilowatt-hours. If you use natural gas for heating in your home, the utility company probably bills you for it in BTUs (short for British thermal units), "therms," or cubic feet of gas used. If you research the energy usage of various countries, tables will often show it to you in "barrels of oil equivalent" or "tons of coal equivalent." If you read a science textbook, you may encounter energy units such as "electron-volts." And so on. This potpourri of units can seem confusing, but it takes only a quick Web search to find the conversion from one energy unit into any other. For example, you'll find that 1 kilowatt-hour is the same as 3.6 million joules, 860 food calories, or 3,412 BTUs, and that one barrel of oil can be used to produce approximately 1,700 kilowatt-hours of energy.

All these units are legitimate, and some are arguably more useful in particular situations than others. However, I find it easiest to think about power in units of watts and energy in units of kilowatt-hours; this allows me to form a mental picture of 1 kilowatt-hour just by visualizing ten 100-watt lightbulbs left on for an hour. Just remember that whenever we speak in terms of watts, kilowatts, megawatts (millions of watts), gigawatts (billions of watts), or terawatts (trillions of watts), we're talking about a *power* requirement, which means the continuous *rate* at which we are using energy; we can then multiply the power requirement by the amount of time that we're using this power to figure out how much total energy is required.

The Energy Problem, Part 1—Finding Enough

We're now ready to look at the problems we face in maintaining our energy economy. Remarkably, if we boil it down to the basics, there are only two major problems: the problem of finding affordable energy sources to provide all the energy we need to run our civilization, and the problem of dealing with the consequences of energy usage, such as the pollution that results from it and the geopolitical implications of importing it versus producing it at home. Let's begin with the problem of finding enough.

The mathematics of the issue is simple. As I've already noted, the United States requires average power of about 3 terawatts, or 3 trillion watts. This is the same as 3 billion kilowatts, which means we use about 3 billion kilowatt-hours of energy every hour. A moment ago I told you that a barrel of oil (a barrel contains 42 gallons) can be used to generate about 1,700 kilowatt-hours of energy. So if we divide our hourly energy needs by the amount of energy we get per barrel of oil, we find that *every hour* the United States requires the equivalent of nearly 1.8 million barrels—which is 74 million gallons—of oil. This happens to be the approximate capacity of one of the modern ships that we call supertankers (and more than the capacity of the infamous *Exxon Valdez*), which means that if all our energy came from imported oil, we'd essentially need a supertanker to arrive in port every single hour. For another point of perspective, consider the 2010 BP/Deepwater Horizon oil spill, which dumped an estimated 200 million gallons of oil into the Gulf of Mexico. As large as this spill was, the lost oil represented barely enough energy to keep our nation going for about three hours.

The United States currently represents about one-fifth of total world energy usage, so the global energy supply problem is five times as large. That is, the world today uses the equivalent of about five supertankers' worth of oil every hour, or 120 supertankers every day, or almost 44,000 every year. In reality, only about 40% of this energy comes from oil (the rest comes mostly from coal and natural gas, with smaller amounts from nuclear energy, hydroelectric dams, and a few "renewable" sources), which means actual oil use is "only" the equivalent of about 18,000 supertanker loads each year. Still, can you even picture the worldwide effort that goes into finding all this oil? It's no wonder that many of the world's largest and most powerful companies are oil companies, and that countries with oil wield far more geopolitical power than their sizes or economies would otherwise indicate.

The magnitude of the problem is even greater when we think about the future. Most of the world's population remains relatively poor, which means that demand for energy will grow as developing nations seek to catch up to the developed world. China, for example, recently surpassed the United States to become the world's largest consumer of energy, yet the Chinese still use only about one-quarter as much energy per person as those of us in the United States. Worldwide, population is expected to grow about another 30% by 2050. Even with much greater energy efficiency, it's likely that global energy demand will double or triple by the end of this century.

As incredible as all this should sound, supply wouldn't necessarily be a problem if we had an unlimited amount of oil. But we don't. Along with coal and natural gas, oil is a *fossil fuel*, meaning it is made from the remains of living organisms that died millions of years ago. Once we use it up, it's gone. It's difficult to determine exactly how much oil still remains in the ground, because most of it is deeply buried in places that are difficult to get to, such as beneath the oceans. That's why you'll see widely varying estimates of when we'll run out of oil. The pessimists think that oil supplies will dwindle greatly in just the next decade or two. The optimists think we may have enough to last several more decades. But I don't see anyone seriously proposing that we have enough oil for it to remain the bedrock of our energy economy throughout this century.

The lesson is clear. Referring again to Mike Bowlin's chapter-opening quote, we are living in "the last days of the age of oil." Unless we find new sources of energy, we cannot maintain the energy economy that we've all become accustomed to. We'll discuss some possible replacements shortly, but first we must look at our second major energy problem.

The Energy Problem, Part 2—
Consequences of Energy Usage

A century ago, major cities were awash in terrible pollution. This pollution, which fouled both the air and the water, sickened tens of thousands of people each year and gave the cities a terrible stench. It was caused primarily by the millions of horses that were used for transportation. At the risk of giving you more details than you really want, I'll point out that besides the obvious problems of horse manure and the associated insects, there were so many horses that disposal of their corpses was also a major contributor to the pollution.

The introduction of automobiles with gasoline-powered engines seemed at the time to be a godsend. It was decades before it began to sink in that these vehicles created serious pollution of their own. Nuclear power seemed like the next savior, offering a promise of cheap, clean energy. But as more nuclear power plants came on line, the difficulty of waste disposal became more clear, and accidents such as those at Three Mile Island and Chernobyl, along with the recent tsunami-inflicted damage to Japan's Fukushima Daiichi Nuclear Power Station, have further dampened public enthusiasm for nuclear energy. Security experts also worry about preventing nuclear power from contributing to weapons proliferation and protecting nuclear power plants against terrorism. Even today's fashionable projects in renewable energy are not without consequences. Wind turbines kill birds and bats, and generate noise and visual clutter that many people don't want in their backyards. Significant waste, some toxic, is incurred in both the manufacture and the disposal of solar panels. Tapping into deep geothermal heat sources may lead to earthquakes. Growing crops for ethanol or other biofuels is energy intensive, making the overall benefits of these fuels unclear, and also leaves less land available on which to grow food for the world's growing population.

The lesson is that no energy source is completely benign, which means that when we compare energy sources, we should be taking into account the costs of their consequences as well as the costs of their production. Unfortunately, as President Nixon noted decades ago in the second quote with which I open the chapter, the prices of energy products do not currently include these costs.

Of course, this is where the mathematics gets tricky, because many of the costs are difficult to quantify. Let's start with the easiest ones. First, there are the health costs of air pollution, which is produced primarily by oil and coal usage. We cannot determine these costs exactly, because almost any health problem that we might attribute to pollution might also have had some other cause; for example, it's almost impossible to determine whether pollution or something else caused any particular asthma attack. Nevertheless, the cost issue can be studied statistically, and if there are more asthma attacks in places with more polluted air, then we can reasonably consider the cost of treating these excess attacks to be a cost of pollution.

Looking at such issues in detail, the National Academy of Sciences recently estimated the direct health costs of air pollution in the United States at about $120 billion per year. Although we should assume a fairly high uncertainty in this number—it could easily be off by 30% or more—the methodology used in the study means it's probably a lowball estimate. The reason is that the study took a conservative approach, counting only those health effects that were most clearly linked to air pollution and that had measurable costs in medical treatment. It did not, for example, consider the cost of the lost productivity of workers staying home but not seeking treatment, or the lost wages of people who died prematurely due to health problems worsened by air pollution. It also left out the health effects of water pollution, which is often a result of contamination incurred during coal mining and oil and gas drilling.

A second cost that is relatively easy to account for is that of securing our oil supply. Because most of our oil is imported, we need to ensure that the supply and the shipping routes are secure against potential aggressors. Again, there will be uncertainty in deciding what fraction of our military costs should be assigned to protecting oil, but numerous studies agree remarkably well in putting the costs of oil security at about $50 billion to $100 billion per year. Note that, like the health cost estimates, these estimates are conservative in that they include only the military costs that are most directly connected to oil. They do not, for example, include the cost of the war and our occupation in Iraq.

What do these health and security costs mean? One way to think about this question is to consider what would happen if we included them in the price of gasoline. To do that, we need only to divide the total costs by the total number of gallons of gasoline we use in the United States each year.[21] Using the estimates above, the health and security costs total to between about $170 and $220 billion per year, and a quick Web search will tell you that we

---

21. By assigning all of the noted health and security costs to gasoline, I am neglecting the fact that some of the health costs are attributable to power plant usage of natural gas and coal. You could try to account for this by estimating the portion of the health costs due to factors besides gasoline usage, but it would not change the general picture that emerges from our results.

consume about 140 billion gallons of gasoline per year in the United States. Therefore, the health and security costs would add somewhere between about $1.20 and $1.60 per gallon to the cost of gasoline, and that's under the conservative assumptions used in the studies that we've discussed. Note that these are real costs that *we as a nation are already paying*; the military cost is funded by taxes, while the health costs are borne primarily through a combination of government spending and higher insurance premiums than we would have otherwise.

Now ask yourself: Which do you think is more fair, the current system, in which we essentially socialize the costs across our population, or charging the costs to end users by adding, say, a $1.50 per gallon tax to gasoline? Reasonable people can disagree, but it's worth noting that many people give answers diametrically opposed to their claimed political philosophies. That is, so-called conservatives generally prefer the current socialized costs in this case, while so-called liberals tend to prefer the option that puts the cost on the users. Personally, I favor a "carbon tax" that would apply not only to gasoline but to all fossil fuels, and I favor it for the economic reason alluded to by Nixon: Only by charging the real costs of energy can we hope to reduce consumption and allow alternative fuels to compete on a level playing field.

If you're surprised by how much the health and security costs would add to the price of gasoline, then hold on to your hat, because the more serious costs are still to come. Let's begin with the cost, in dollars and lives, of oil imports through which we effectively transfer hundreds of billions of dollars to nations that don't share our values. Without oil income, does anyone seriously believe that Iran would be on the verge of developing nuclear weapons, that Hugo Chavez would still be running Venezuela, or that we would be supporting a regime in Saudi Arabia that doesn't even allow women to drive? And don't forget that most of Al Qaeda's money is ultimately oil money, as is most of the money going to other terrorist groups such as Hamas and Hezbollah. It's extremely difficult to put a price on the insanity of handing our money over to such people, but I suspect that price is far higher than the $1.50 or so per gallon of gas that we've discussed so far.

We also have not yet accounted for the environmental costs of losing mountaintops to coal production, losing lakes to acid rain, or destroying ecosystems through oil spills. Nor have we considered the future liabilities associated with using up a limited resource. After all, we always use the most accessible supplies of coal and oil, which means that any oil and coal that our children and grandchildren use will be more difficult and more expensive to extract. Again, there is no way to place exact dollar values on these downsides, but I suspect that if we include everything we've discussed so far, the real price of fossil fuels is at least double or triple what we currently pay.

In the long term, however, all of these costs may pale compared to the most difficult cost of all to quantify: the cost of climate change brought on by global warming. And that brings us to a question I raised back in Chapter 1: Given the wide range of opinions that you hear about global warming in the media, how can you decide whether it's a serious threat or some type of hoax?

Although you probably think of global warming as a science issue, it involves a great deal of math in the form of measurements, calculations, statistics, and the building of mathematical models of the climate. So let's treat the question as a math research problem. The basic claim made by the scientists who talk about global warming is that the burning of fossil fuels releases *greenhouse gases*—most notably carbon dioxide—that are trapping more heat in the atmosphere, causing our world to warm up. Our research problem is to test this claim.

The first step is to find out whether carbon dioxide and other greenhouse gases really can trap heat and, if so, how much. This question can be explored in the laboratory by, for example, putting greenhouse gases in tubes and shining light (especially infrared light) through them, so that we can measure how much light of various wavelengths they absorb. These types of measurements have been made many times (beginning more than 150 years ago), so the precise heat-trapping effects of carbon dioxide and of other greenhouse gases are very well known, and not subject to any dispute.

Next, we need to figure out whether the heat-trapping effects of greenhouse gases can really make a planet warmer than it would be otherwise. To do this, we need to think about the major factors that determine a planet's average temperature, and there are only three: its distance from the Sun, the fraction of the incoming sunlight that its surface absorbs (the rest is reflected back into space), and the amount of heat trapped by greenhouse gases in its atmosphere. There are relatively simple equations that take into account all three factors, and we can be confident that these equations are correct, because they successfully reproduce the observed average temperatures of all the planets. Moreover, by comparing these observed temperatures to the temperatures expected in the absence of greenhouse gases, we find that more greenhouse gases mean more warming. Our planetary neighbors vividly demonstrate this fact. Mercury, which has no atmosphere, has the average temperature we expect without greenhouse gases. Mars has a very thin carbon dioxide atmosphere that gives it a fairly weak greenhouse effect, making the planet about 11°F warmer than it would be otherwise. Venus, which has an extremely dense atmosphere containing nearly 200,000 times as much carbon dioxide as Earth's atmosphere, has a correspondingly extreme greenhouse effect that makes its surface about 850°F hotter than it would be otherwise—hot enough to melt lead.

Earth is the lucky intermediate case. Although our atmosphere is composed mostly of nitrogen and oxygen, neither of which have any heat-trapping effects, it contains just enough of the greenhouse gases carbon dioxide, methane, and water vapor to raise the temperature about 55°F higher than it would be otherwise. Since the actual average temperature is close to 60°F, we conclude that our planet would be frozen over without the naturally occurring greenhouse effect. From that standpoint, the greenhouse effect is a very good thing, because our lives would not be possible without it. Just keep in mind that the case of Venus offers proof that it's possible to have too much of this good thing.

Having established that greenhouse gases really do trap heat and increase planetary temperatures, we turn to the claim that the burning of fossil fuels is adding to the greenhouse gas concentration of our atmosphere. Again, this is a measurable effect. Since the 1950s, scientists have been making direct measurements of the amount of carbon dioxide in the atmosphere. For earlier times, scientists can measure the amount of carbon dioxide trapped in bubbles in ice cores drilled from glaciers, the age of which can be determined in much the same way as the age of tree rings; these data allow us to measure the carbon dioxide concentration going back almost a million years. Figure 14 shows both data sets, with the concentration in units of parts per million, which means the number of carbon dioxide molecules in each one million molecules of atmosphere. I want you to notice three crucial facts. First, although carbon dioxide plays a huge role in Earth's temperature, it is only a trace gas in our atmosphere; for example, a concentration of 300 parts per million represents only 0.03% of the atmosphere. Second, the atmospheric concentration of carbon dioxide has increased nearly 25% in just the past five

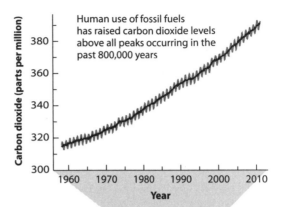

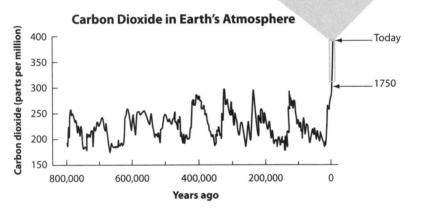

Figure 14. The lower graph shows ice core data for the carbon dioxide concentration in Earth's atmosphere over the past 800,000 years (from the European Project for Ice Coring in Antarctica). The zoom-in shows the concentration measured directly over Mauna Loa since the 1950s (data from the National Oceanic and Atmospheric Administration).

decades, from about 315 parts per million in the late 1950s to 390 parts per million in 2011, and it is continuing to rise at a rate of about 2 to 3 parts per million each year. Third, the carbon dioxide concentration is now substantially higher than it has been any time in the 800,000 years for which data are available, which means we are entering uncharted territory in terms of understanding how it will affect our planet.

Could the carbon dioxide concentration be increasing naturally, rather than due to the burning of fossil fuels? This, too, can be checked, because carbon released by fossil fuels has a slightly different chemical makeup (in its ratio of isotopes) than carbon from other sources. The results leave no room for doubt that most of the new carbon dioxide entering the atmosphere comes from fossil fuels.

Let's summarize what we've found in investigating the claim of global warming:

- The heat-trapping effects of carbon dioxide and other greenhouse gases have been directly measured in the laboratory.
- Calculations show that accounting for these effects is necessary to explaining observed planetary temperatures, and both observations and calculations show that more greenhouse gas means a hotter planet.
- Measurements show that the concentration of carbon dioxide in Earth's atmosphere is rapidly rising, and that this carbon dioxide comes from the burning of fossil fuels.

Perhaps it's not quite as obvious as $2 + 2 = 4$, but there seems to be a clear, mathematically based truth here: The burning of fossil fuels does indeed release gases into the atmosphere that cause global warming.

The only way left to get around the conclusion that global warming is a serious threat is to imagine that Earth has some feedback mechanism that can somehow counteract the warming effects of the greenhouse gases. To study this possibility, scientists create mathematical models that use equations to represent thousands of known interactions that occur in Earth's climate system. Then, using measured data from the past and present as inputs, they run these models on supercomputers to make simulations of Earth's climate. The models can never be perfect, because there will always be interactions that they cannot fully account for. But we can test whether the models are on the right track by seeing how well the simulated climates in the computer match Earth's actual climate. Today's models provide remarkably close matches to the real climate, making it unlikely that they are missing any major component of the climate system. Indeed, if there were any missing component that could mitigate global warming, it would have to have some rather odd properties. For example, it would have to be strong enough to counter the future effects of a buildup of carbon dioxide, yet weak enough so that its absence wouldn't already be causing the models to give a poor match to the actual climate. We can't rule out the possibility that such a component exists, but I wouldn't recommend betting the future of our civilization on it.

Can we account for the costs of global warming in the price of energy? To do so, we would have to quantify the risks associated with global warming. The main difficulty is that, because our climate models are not perfect, we have no way to know whether their future predictions will prove accurate until the future arrives. That said, it's still worth looking at what the models suggest. The predictions vary depending on the precise model used, but generally speaking, all of the models predict that global warming will lead to rising sea level; more frequent and more severe storms; and regional changes in climate that could lead to reduced agricultural production, increased loss of life and property due to disasters such as wildfires and flooding, and the collapse of localized land and ocean ecosystems. The more carbon dioxide we continue to add to the atmosphere, the worse those effects will be, and the more rapidly they will be upon us.

Worse, if we do enough damage, we know of additional factors that could enter the mix and exacerbate the problems. For example, vast amounts of the greenhouse gas methane are currently trapped in Arctic regions; if this methane is released, it will amplify the global warming trend. Another possibility is that warming could lead to changes in the ocean currents that moderate many coastal climates. Perhaps worst of all, the melting of ice at the polar caps could lead to significant increases in sea level. Together, Antarctica and Greenland contain so much ice that, if it all melted, sea level would rise by more than 200 feet. Although scientists think it would take many centuries to melt this much ice even in the worst case, the melting of just a small fraction of it could cause severe effects. For example, many scientists think it likely that enough melting will occur to increase sea level by several feet during this century, which would lead to much more severe coastal flooding and would likely wipe out the homes of hundreds of millions of people around the world. The economic costs of any or all of these consequences of global warming are hard to quantify, but it is safe to assume that they will far exceed the costs of, for example, securing our oil supply.

The bottom line is that the real cost of fossil fuels is far higher than the cost we have been paying. We've found that these fuels are underpriced by at least the equivalent of about $1.50 per gallon, even if we account only for the easily quantifiable costs, and by a factor of two or three if we add in other known geopolitical and environmental factors. The effects of global warming could make the actual cost much higher still; in the extreme case, the price could be economic devastation. You can form your own opinion, but mine is clear: We simply cannot afford to maintain our energy status quo.

Potential Solutions

If, as I believe, the status quo is a recipe for disaster, then we have no choice but to change our energy portfolio, with the goal of finding options that provide the needed energy supply while minimizing the consequences. The path to this goal, however, is unclear and subject to great debate. Although it will take us somewhat away from our usual focus on quantitative issues, I'd be remiss if I didn't at least summarize the major options available to us. As you've probably come to expect, I'll also interject some of my personal opinions about which options are best.

Reduced Demand

The cheapest and easiest way to make sure we have enough energy supply is to reduce energy demand. If we could cut our energy use in half, for example, then we'd need to find only half as much supply, and we'd also cut in half all the consequences of using energy.

There are two basic ways to reduce demand: doing without some of the comforts we've become accustomed to, or improving the efficiency of our energy-using devices. I have friends and neighbors who have done remarkably well at the first strategy through such techniques as walking or biking almost everywhere they go, hanging out clothes to dry instead of using a dryer, and lowering their thermostat in winter. But this type of energy savings requires great dedication; I like to think that I'm doing my part to help solve the problem, but I still often drive even when I have other options, and I use a clothes dryer and keep the winter thermostat at a comfortable 69°F. So the realist in me says that if we really want to make strides in reducing demand, we'll need to make them on the efficiency side.

In fact, it's fairly easy to see ways in which efficiency could cut demand by half or more. Hybrid cars such as the Prius get nearly double the gas mileage of the average car driven in America and therefore offer a clear path to cutting gasoline usage almost in half. Electric cars might do even better; electric motors are generally more efficient than gasoline engines, so if the power plant supplying the electricity is also efficient, then the total energy required per mile of driving can be reduced substantially. Other electrical savings can come from replacing old appliances with more efficient models and by changing old lightbulbs to newer CFLs or LEDs. The energy needed for heating and cooling can be cut substantially simply by using better insulation and better windows in buildings.

It's worth noting that most of the technology needed to make substantial cuts in demand already exists. The only reason it isn't widely used is the startup cost of change—such as the cost of buying a more fuel-efficient car, changing out lightbulbs, or adding insulation. Policy makers can address this problem by mandating greater efficiency. But an even better idea would be to let the market do it: If we charged more realistic prices for our current energy sources (by adding a carbon tax that would help account for the real costs of fossil fuels), people would quickly realize the monetary value of conservation and efficiency.

Overall, reducing demand is a "no-brainer": It helps both the supply and the consequence problems, while saving money at the same time. But it won't be enough by itself. Even if we double our global energy efficiency, the growing demand from developing nations is likely to make up for those savings. We therefore need to find new ways to produce energy as well as to save it.

Clean Coal

We may be in the last days of the age of oil, but coal is a very different story. Estimates suggest that there is enough coal to satisfy global energy demand for at least a century, and possibly for two or three centuries. The United States has the world's largest supply of coal, so a switch to coal means we'd no longer have to import energy. Moreover, changing from oil to coal would require only minimal changes to our energy infrastructure, since we already generate much of our electricity with coal-fired power plants, and it is possible to "gasify" coal to produce gasoline that would work in current automobiles. The downsides of coal come on the consequences side. It is the dirtiest of the fossil fuels, both in the visible effects of mining it and in the toxic pollutants it releases. It also releases more carbon dioxide per kilowatt-hour of energy produced than either oil or natural gas. As a result, we can only afford the real costs of coal if we can find ways to clean it up. Can we?

Technologies to mitigate the environmental effects of coal mining and to reduce the pollutants coming from coal burning already exist, though their high costs mean they are not yet used as widely as they could be, particularly in China and other developing nations. But the biggest problem with coal is its carbon dioxide emissions. Nothing can be done to prevent the production of carbon dioxide when you burn coal; therefore, hopes for "clean coal" rely on the prospect of *carbon sequestration*—finding a way to lock away the carbon dioxide produced from coal before it can enter the atmosphere. There are two basic approaches to carbon sequestration. One would inject the carbon dioxide deep into the ground where it cannot escape; the other would use chemical or biochemical processes to lock the carbon dioxide into some solid structure, much as nature can lock it into the rock known as limestone. Several test projects are currently under way, but neither approach has yet been proven. As a result, we still do not know whether clean coal is really possible.

Even if it does prove possible, clean coal will still have at least three significant negatives. First, no matter how much mitigation we do, mining will inevitably have serious environmental consequences, and the more we mine, the worse those will become. Second, there will always be an incentive for those producing power from coal to cheat and burn it "dirty," because that will be cheaper than adding the clean technologies. While developed nations may be able to prevent such cheating through strict enforcement mechanisms, the rampant corruption in many poorer countries will make it much more difficult to count on clean coal in developing economies. Third, even if coal supplies last a century or more, they are still limited, so clean coal is at best an interim solution to civilization's energy problems.

My personal opinion is that we should continue to explore clean-coal possibilities. If they pan out, clean coal may be a great short-term solution to our immediate energy problems. But for the long term, we need to look in other directions.

Natural Gas

The third fossil fuel is natural gas, and a lot of people have been actively lobbying for greater use of it, primarily because it burns cleaner and emits less carbon dioxide per unit of energy than either oil or coal, and because the United States has a decent supply of it. However, while it emits *less* carbon dioxide than other fossil fuels, it's still a significant source of the gas, which means it can only slow the growth of our carbon dioxide problem, not eliminate it. Moreover, if we use it as our major source of energy, we at best have only enough natural gas to last a few decades, and recent studies suggest numerous other risks associated with the process of extracting natural gas from the ground, all of which add to its true costs. Like clean coal, natural gas has at least some potential to help solve our short-term problems, but we're still in need of long-term solutions.

Renewable Energy

The most fashionable alternatives to fossil fuels are renewable energy sources, such as wind, solar, geothermal, hydroelectric, and biofuels. As we've discussed, these are not perfect, but they have far less serious consequences than fossil fuels. The debate about renewables therefore focuses on how much energy they can realistically provide and whether they are cost effective.

The news is good on the supply side. Current global power demand is about 15 terawatts. The wind carries more than ten times that much power around the globe, and accessible wind sources are estimated to be enough to supply about 20 terawatts. Solar is even more promising: Sunlight supplies Earth with continuous power of about 100,000 terawatts, so solar could more than meet all our power needs if we collected just 0.02% of the incoming sunlight. The promise of biofuels from crops is less clear, due to the problems noted earlier—growing the crops is energy intensive and removes farmland from food production—but biofuels could still help meet at least some of our energy demand. The same is true for hydroelectric, which holds great promise, but in some forms (such as dams) has severe ecological consequences of its own. Geothermal is an interesting case: While deep geothermal sources have posed worries of earthquake risks, there's a more readily available type of geothermal energy that is already being used to heat and cool thousands of homes and business. It works simply because the underground temperature remains nearly constant year-round, which means that heat can be extracted from the ground in the winter and dumped into it in the summer.

Let's turn to the question of cost effectiveness. It's not easy to compare the prices of renewables to those of fossil fuels because a wind turbine or solar panel, for example, can continue to generate energy for a long time, while fossil fuels must be continually resupplied. You therefore have to pick a time horizon over which to spread the installation costs of the renewables, then estimate how much energy you'll get from them over that period. Most estimates suggest that the price of energy from renewable sources such as wind, solar, and geothermal is now less than double what we are paying for fossil fuels, which means it is probably lower than the true costs (including the costs of the consequences) of fossil fuels.

The tremendous potential and low costs of renewables have led many environmentalists to envision a rapid transition from our current fossil fuel–based portfolio to a renewable-energy portfolio. However, there are some practical problems that may make this transition much more difficult than it sounds. For example, our current energy "grid"—the system of power cables and power stations that transports electricity—would need to be substantially rebuilt to use power sources that are intermittent rather than constant. We'd also have to solve the problem of storing energy for nighttime, cloudy days, and times when the wind is weak. Moreover, despite their global environmental advantages, large-scale wind and solar projects are already encountering strong opposition from people who worry about their local impacts.

My own guess is that renewables will play an important role in our solutions, but they won't be a panacea. If we priced energy fairly by adding a carbon tax to the cost of fossil fuels, I think that nearly all of us would have solar panels on our roofs, and we'd find a way to panel all the commercial buildings and parking garages, too. I also suspect that geothermal heating and cooling would become the norm rather than a rarity. But for all this to become a reality, we're going to have to convince our politicians to charge us the true costs of energy.

Nuclear Power

Perhaps the most controversial of the potential solutions is nuclear power, which already supplies a substantial fraction of the global energy supply. Note that this nuclear power is *not* the nuclear fusion that we discussed at the beginning of the chapter, in which hydrogen is fused to make helium. Instead, it is nuclear *fission*, in which large atoms of uranium or plutonium are split apart, releasing energy in the process. Proponents of nuclear power point out that it is relatively cheap and produces no greenhouse gases. In that sense, it could be our best hope of solving the problem of global warming. Opponents cite the problems of disposing of radioactive waste, of protecting against terrorism and weapons proliferation, and of the danger posed by radiation-releasing accidents, with this last drawing special attention since the earthquake-induced nuclear accident in Japan.

The good news is that there are a number of new technologies for building nuclear power plants that can dramatically reduce the risk of radiation-releasing accidents. (Of course, most of these technologies help only future power plants, so we must separately consider how to reduce the danger at existing ones.) The terrorism and weapons proliferation problems are probably also tractable, though only with strong international cooperation. That leaves the problem of waste disposal. Nuclear waste remains dangerous for tens of thousands of years. While some new technologies promise to reduce the total amount of waste, we ultimately will need to find ways to keep the remaining waste out of reach for a thousand generations. The only practical solution is some type of permanent burial, but that has proven to be very difficult. For example, the United States has already invested billions of dollars in creating a nuclear waste burial facility at Yucca Mountain in Nevada, but concerns about whether the site really offers "permanent" burial have so far prevented its use.

We are left with a clear trade-off on nuclear power: It has known and very real dangers, but could also save our planet from the threat of global warming. In my view, the dangers make nuclear power the option of last resort . . . but since we've reached the end of my list of major energy supplies, and have not yet found a full solution to the energy problem, I'm afraid we need a last resort. I therefore think we should go full bore into developing as much nuclear power as we can (using the new, safer technologies), because even under the best of scenarios, it will take decades for it to come on line. With luck, the combination of nuclear power, renewable sources, greater energy efficiency, and perhaps some clean coal and gas will allow us to stop global warming before it is too late, and buy us time to find a better long-term source of energy.

Game-Changing Technologies

What would a better long-term solution look like? There are a number of possibilities on the horizon, starting with nuclear fusion. Our kitchen-sink example showed the tremendous potential of fusion energy. Perhaps equally astonishing, fusion arguably has smaller consequences than almost any other energy source. The process itself turns hydrogen into helium, a harmless (and useful) gas; no greenhouse gases are produced. The amount of water needed to supply the hydrogen is so small as to have no significant impact. Fusion reactors would not pose a danger of major accident; unlike fission, which can spiral out of control if there is a reactor failure, fusion automatically turns itself off if the reactor fails in any way. Moreover, we know that fusion energy is possible, at least in principle, since it is the energy source of the Sun and stars. As Carl Sagan said: "Every time you look up at the sky, every one of those points of light is a reminder that fusion power is extractable from hydrogen and other light elements." The only significant environmental consequence of fusion is that neutrons produced as byproducts would gradually make the walls of a fusion reactor radioactive. This would create a waste disposal problem, but one far smaller than the one created by current nuclear power plants.

The problem is that, despite decades of effort, we have not yet developed the technology for commercial fusion reactors. (We know how to release fusion energy in sudden bursts—that is what a thermonuclear bomb does—but we don't know how to release it at a gradual pace without the process simply turning itself off.) A new international effort, known as ITER, is hoping to bring the technology closer to reality, but its construction has only just begun. Some scientists doubt that we can realistically obtain energy from fusion at all, and the track record isn't encouraging; it's become a running joke among scientists that fusion will forever be "a couple decades away," because those working on fusion have been saying that for more than a half century. But given that we know fusion energy is possible in principle, I personally suspect that we could achieve it, if we made a great enough effort.

Another potential game-changing technology, and one of my personal favorites, is to launch solar panels into space. If we put the panels in a high orbit, they'd be in continuous sunlight, with no night or clouds to interrupt their power collection, and they could supply many times the world's current power needs without any noticeable impact on the amount of sunlight reaching the ground. The solar energy could be beamed down to Earth in the form of concentrated light, collected at ground stations dispersed around the world. This would, of course, pose a danger to any birds or airplanes flying through the beams, but airplanes could avoid the beam paths, and the birds, well . . . given that there'd be no pollution, no greenhouse gases, and no mining or drilling, the total environmental impact would be far lower than that of our current energy sources. Moreover, most of the technology for solar power from space already exists; the primary difficulty is the high cost of launching thousands of square kilometers of solar collectors into space, and of setting up the ground infrastructure for collecting and distributing the beamed-down energy.

Biotechnology also provides a possible avenue to an energy solution. Craig Venter, whose research group was the first to sequence the human genome, is working on engineering bacteria that would ingest carbon dioxide from the atmosphere and convert it into fuel. Other groups are working on making biofuels from algae, a process that avoids many of the problems of crop-based biofuels, such as ethanol from corn.

The list goes on, but keep in mind that while any one of these might become the game-changing technology that saves our future and our planet, it's also possible that none of them will succeed.

Geoengineering

Imagine that, a couple of decades from now, global warming is rapidly worsening, and no new technologies have come along to allow us to stop it while still providing for our energy supply. What then? A few people have been pondering schemes that might somehow counter the planetwide warming effects of carbon dioxide, schemes that go under the moniker of *geoengineering*. For example, some people have proposed seeding the atmosphere with aerosols that would reflect sunlight back to space, or even deploying giant sunshades in space. The problem is that most of these ideas do not actually reduce the amount of carbon dioxide in the atmosphere, and therefore suffer from at least three major drawbacks. First, as long as we continue to add carbon dioxide to the atmosphere, some of it will dissolve in the oceans, where it causes *ocean acidification* that can destroy ocean ecosystems; this problem could potentially prove to be just as devastating as the warming of our planet. Second, most geoengineering proposals require active maintenance; for example, the aerosol idea requires continually putting more aerosols in the atmosphere to replace those that rain out. If the maintenance ever failed— whether now or centuries from now—the global warming problem would immediately return, and it could be far worse if we'd continued to burn fossil fuels in the interim. Third, geoengineering poses risks that we cannot predict well, because it introduces global climate factors that do not exist naturally and therefore are difficult to account for in models; I doubt that we could ever be truly confident that the cure wouldn't be worse than the disease.

Given these drawbacks, I believe there's only one type of geoengineering worthy of serious consideration: trying to scrub carbon dioxide from the atmosphere, incorporating it into rock or other solid form that can then be stored away safely. If we could do that, then we could potentially stop or even reverse global warming regardless of our fuel source. Unfortunately, while a few ideas are being researched (including Venter's idea for bacteria that would turn carbon dioxide into fuel), no one yet knows how to scrub carbon dioxide in this way. Moreover, even this technology would require great care to avoid inadvertent consequences, such as removing so much carbon dioxide that we ended up cooling our planet.

The Bottom Line

There are many possible ways that we might address our energy supply and consequence problems, but unless one of the game-changing technologies comes through, no single option is likely to be enough. My own recommendation is simple. We should do everything we can to increase energy efficiency and the use of renewable resources. We should plan for a dramatic increase in our use of nuclear power. And we should invest heavily in searching for game-changing solutions. How can we make all this happen? Equally simple: let the market encourage human ingenuity by making sure that what we pay for energy more closely matches its true costs, something we can do by applying a substantial carbon tax[22] to fossil fuels. This will immediately make alternative technologies competitive, and the money raised by the tax can be invested in research for even greater possibilities.

Math for Life

We rarely think of energy as a mathematical issue, but I hope I've shown you in this chapter that the way we deal with energy does indeed require mathematical thinking. Sadly, we as a nation don't seem to be doing very well at this kind of thinking. Let's briefly revisit three examples from this chapter. First, consider fusion. Knowing that your kitchen sink could in principle supply enough hydrogen to power our entire nation, you might expect that we'd be spending a lot of money to find out whether we can really make it work. In fact, current spending on fusion research in the United States is about $400 million per year (including our contribution to ITER), which is barely $1 per person—far less than we spend on soda, candy, pizza, movies, or buying gas for short trips on which we could have easily walked.

Next, look back at our discussion of global warming. Note that none of the measurements or calculations underlying the basic concept of global warming are subject to any scientific dispute at all; the only debatable part is the modeling, and even that looks pretty solid. So how can so many politicians and pundits claim that global warming is some kind of hoax? Unless they're simply not paying attention, we're forced to conclude that they're very bad at math. Many of their pronouncements indicate that they do indeed have a math problem. For example, when the East Coast suffered a cold winter in early 2010, we heard many of these folks proclaiming it as "proof" that

---

22. While there has been little serious effort in Congress to implement a carbon tax, there has been some effort to introduce "cap and trade," a system that places a cap on total carbon dioxide emissions while allowing companies to trade carbon dioxide credits in a way that should theoretically encourage them to reduce emissions. My own opinion, since you didn't ask: This system is overly complex, still hides the true cost of fossil fuels from view, and in the end is just a tax by another name. Its only potential advantage over a straight tax is that the cap can be adjusted downward with time, but I believe that a carbon tax that reflects the true cost of fossil fuels would encourage enough market innovation to bring total emissions down even more rapidly, and therefore is a much better way to go.

global warming isn't real. But anyone who understands even minimal statistics would know that localized conditions along the East Coast tell us nothing about what is happening globally (in fact, globally averaged, this was one of the warmest periods in recorded history), and that a single season does not tell us much about long-term trends.

Finally, and most importantly, there's the conflict between our supposedly free-market views and the fact that we socialize the costs of our current energy use. This suggests we don't even understand enough math to apply our own philosophies consistently. Clearly, if we are going to solve our energy problems, we need to get a lot better at math. Let's hope that we can do this, because it is critical to our survival. To help make this point, I'll conclude with a wonderful quotation that, although spoken in a different context, applies just as well to our energy future:

> We are now faced with the fact, my friends, that tomorrow is today. We are confronted with the fierce urgency of now. In this unfolding conundrum of life and history, there is such a thing as being too late.
> — **Martin Luther King Jr.**

8

The Math of Political Polarization

These days, almost every congressional district is drawn by the ruling party with computer-driven precision to ensure that a clear majority of Democrats or Republicans reside within its borders. Indeed, it's not a stretch to say that most voters no longer choose their representatives; instead, representatives choose their voters.
— **Barack Obama, in *The Audacity of Hope***

We are going to draw the lines so that Republicans will be in oblivion in the state of New York for the next 20 years.
— **New York State Senator Malcolm Smith, speaking about redistricting**

We'll start this time with two closely related questions.

Question 1: The 2008 and 2010 elections both led to changes in governing power, with Democrats gaining the presidency and the Senate in 2008 (they already had control in the House) and Republicans gaining the House in 2010. Since there are 435 members of the House of Representatives, these two elections involved a total of 870 individual House races. What would you guess was the average (mean) margin of victory among those 870 House races?

Answer choices:
a. Less than 3 percentage points
b. 3 to 7 percentage points
c. 8 to 15 percentage points
d. 16 to 29 percentage points
e. More than 30 percentage points

Question 2: Suppose that in the next congressional election, a state with 8 seats in the House of Representatives has voters that are evenly divided in their preference for Republicans and Democrats; that is, suppose that exactly 50% of the voters will vote for a Republican and the other 50% for a Democrat. How many of the 8 House seats would you expect to be won by a Republican?

Answer choices:
a. 4
b. Either 3, 4, or 5
c. Either 2, 3, 4, 5, or 6
d. Either 1, 2, 3, 4, 5, 6, or 7
e. All the elections will be ties

Let's start with the first question. I've posed this one to many people, and nearly all of them guess that the answer is A. It's a sensible choice; after all, the fact that power changed hands in both elections suggests a fairly closely divided electorate. Polls confirm this fact: Americans are almost evenly split, with roughly one-third identifying as Republicans, one-third as Democrats, and one-third as independent of either major party. With a closely divided electorate, you might expect most elections to be close. But A is not the correct answer.

You can find all the House election results for 2008 and 2010 online, and if you copy them into a spreadsheet, then you can easily calculate the average (mean) margin of victory for the 870 races. The result is stunning: The correct answer to our question is E, because the mean margin of victory was nearly 34 percentage points. In other words, the average victor had a vote total about 34 percentage points higher than the average loser, which means that the average vote was about 67% to the winner and 33% to the loser, for a 2-to-1 margin of victory. In fact, many incumbents have such a clear path to victory that no serious candidates even bother to challenge them.

You can get a more vivid sense of the disconnect between voters' overall party preferences and the way that House races are decided by comparing the results of presidential elections to those of House elections. Table 2 shows these results for the past five presidential election cycles. Notice that the presidential races are quite close, and the winning party has switched back

Table 2. Presidential and House Election Comparison

Year	President's Popular Vote Margin of Victory* (percentage points) with winning party in parentheses	House of Representatives Mean Margin of Victory (percentage points)	Percentage of House Incumbents Winning Reelection
1992	5.6 (D)	30.5	88
1996	8.5 (D)	30.4	94
2000	−0.5* (R)	39.9	98
2004	2.4 (R)	40.5	98
2008	7.3 (D)	37	94

* The margin is negative because the electoral vote winner lost the popular vote.

Sources: Fairvote.org, Opensecrets.org

and forth, just as we might expect for a closely divided electorate. In contrast, the mean margin of victory in House races is enormous; as the last column shows, one consequence of this fact is that nearly all incumbents are reelected. (Even in the dramatic "Republican tide" of the 2010 elections, about 85% of the incumbents who ran for reelection won.) The high reelection rates of incumbents are even more astonishing when you consider how they contrast with public approval of the job Congress is doing. In 2008, for example, polls showed the House with a less than 20% public approval rate overall, yet we reelected 94% of the incumbent members who ran.

Given that we have such a closely divided electorate, you might wonder how the average margin of victory for House races can be so huge. The reason is directly traceable to our second question at the start of this chapter, for which the correct answer is D. This answer also surprises most people, because with a 50:50 split in voter preferences, you might expect the congressional delegation to be split evenly as well, or for all the elections to be very close to being ties. However, depending on the geographical distribution of voters and who draws the district boundaries, the actual result can go as much as 7 to 1 in favor of one or the other party. To see why, let's take some sample numbers.

Suppose that the state has 500,000 voters in each of its 8 House districts, for a total of 4 million voters in the state. Because we've assumed evenly divided preferences, we expect that 2 million voters will vote for a Republican and 2 million will vote for a Democrat. Let's further suppose that, as is often the case, the Democrats and Republicans tend to be concentrated in certain geographical regions; for example, it's common for urban areas to have a higher concentration of Democratic voters and suburban and rural areas to be more Republican.

Now, imagine that the state legislature draws the district boundaries, and that it is controlled by the Democrats at the time. Although it's not quite realistic, let's take an extreme case and assume that the legislature draws a district in which all 500,000 voters are Republican, so that the Republican candidate will win 100% of the vote in this district. For the rest of the state, then, all 2 million Democratic voters remain, while only 1.5 million Republican voters are still available. If the districts can be drawn to split these voters evenly among them, then the Democrats will win every one of the other 7 districts by a margin of 2 to 1.5, which is the same as 4 to 3, which is the same as about 57% to 43%.

Thus, even though the statewide total showed 50% of the vote going to a Republican and 50% going to a Democrat, the Democrats end up winning 7 out of the 8 districts. Moreover, none of the elections are even close. The Democrats' margin of victory is 14 percentage points in their 7 wins, while the lone Republican victory is by a margin of 100 percentage points. The average (mean) margin of victory, then, is close to 25 percentage points.

While packing one district with 100% Republicans isn't really possible, you can see the basic principle that is at work. For a real-world example, consider what happened in Texas in 2002 and 2004. Both elections had very similar statewide results, with Republican candidates getting about 55% of the vote and Democratic candidates getting about 43% (the rest went to third-party candidates). In 2002, using district boundaries that had been drawn by Democrats, the Democrats won 17 of Texas's 32 House seats, leaving 15 to the Republicans; notice that the Democrats won a majority of the seats despite the fact that the Republicans got the majority of the votes. In 2004, using new district boundaries drawn by Republicans, the Republicans won 21 seats to the Democrats' 11, giving them a House seat majority significantly larger than their vote majority. In other words, the mere redrawing of district boundaries dramatically changed the distribution of the House seats, even with essentially no change in the electorate's preferences.

You can probably now understand why I chose the two particular quotes that open this chapter. Malcolm Smith's party, the Democrats, controlled the drawing of district boundaries for New York, and he intended to make sure it drew them in a way that would maximize Democratic power. Obama's quote explains the details behind such power grabs: By using sophisticated techniques to concentrate voters of different preferences into particular districts, the party in power is essentially able to choose who votes where, and thereby to make it more likely that voters will keep it in power.

Political Polarization

Have you ever noticed that most people you know seem pretty middle-of-the-road in their political views, yet many of our politicians seem quite extreme? The questions we've just discussed go a long way toward explaining why.

In an election that is likely to be close, both parties have an incentive to nominate candidates who will appeal to the broad political middle. But in an election that one party is almost guaranteed to win, the real contest occurs in the primary rather than in the general election. Primaries tend to draw much smaller numbers of voters than general elections, and the most highly partisan voters have an extra incentive to show up for a primary if they believe that their candidate is almost certain to win the general election. As a result, noncompetitive districts tend to elect representatives with more extreme partisan views. Moreover, because this effect tends to bring out the partisan voters in the primary, it can also affect nominations for Senate and other statewide offices, which is a major reason why we've seen a trend toward more highly partisan candidates for all offices.

What does all this have to do with math? I hope it's now obvious, but *the entire election process is math.* For House elections, the basic process involves collecting data about voter preferences and then trying different district boundaries until you come up with a set that maximizes your own party's political advantages. Let's investigate how it works in a bit more detail.

Apportionment and Redistricting

In principle, the process of *redistricting*—the redrawing of boundaries for House districts—is supposed to ensure that people are fairly represented by their members of Congress. A quick constitutional refresher may help. The U.S. Constitution established the legislative (law-making) branch of government with two bodies: the Senate and the House of Representatives. Each state gets two senators regardless of its size, but the number of House members depends on each state's population. Because populations tend to change over time, the Constitution directs Congress to conduct a census every ten years, and to reapportion House seats based on the census results.

The reapportioning of House seats among the states is an interesting mathematical problem in and of itself. To understand the problem, consider the current case. The 2010 census found a U.S. population of about 309 million people, and the House currently has 435 seats. If we divide the population by the number of seats, we find that on average, each House member should represent about 710,000 people. However, because state populations are not all multiples of 710,000, there's no perfect way to divide up the House seats. For example, suppose there were a state with a population of 1 million. If you gave that state just one House seat, then its population would be underrepresented in the House compared to those of most other states. But if you gave the state two House seats, then it would have one House member for every 500,000 people—which means it would be overrepresented in the House, and some other state would be underrepresented as a result.

Because House seats can't be perfectly divided among the states, there has to be some way of deciding how many seats go to each state. The country's founding fathers quickly turned to mathematics for this decision process. Alexander Hamilton and Thomas Jefferson each proposed a different mathematical algorithm for apportionment. Both the House and the Senate accepted Hamilton's proposal in 1791, but President George Washington vetoed it (in the first presidential veto in U.S. history). As a result, Congress ultimately adopted and used Jefferson's method. Perhaps not coincidentally, Jefferson's method allotted his home state of Virginia one more seat than it would have received under Hamilton's method (the extra seat came at the expense of Delaware).[23] Worth noting: While Hamilton and Jefferson are

---

23. Hamilton's method was not forgotten, however, as it was used for apportionment between 1850 and 1900.

best remembered as politicians, you can see that they were also pretty good with mathematics.

The fact that different methods can lead to different representation for the states means that apportionment is a very important issue, and mathematicians have therefore studied it in depth. It turns out that, under any reasonable set of standards of fairness, there is no single method that will always work best. Numerous different apportionment algorithms have been proposed over time, and several have been used at different times during U.S. history. Fortunately, different methods yield only slightly different results; even in the most extreme cases, only a handful of states will receive or lose an extra seat by choosing one method over another. Since 1940, Congress has apportioned seats according to the "Hill-Huntington method," which was first proposed in 1911 by the chief statistician of the Census Bureau, Joseph Hill, and Harvard mathematician Edward Huntington.

Apportionment sometimes still causes controversy, especially in states that stand to lose seats after a new census. However, because Congress has stuck with the Hill-Huntington method for so long now, and because the technique is purely mathematical, there's really no way for one side or the other to take political advantage of it. Redistricting is a very different story.

Once Congress has apportioned seats to the states, the Constitution is silent on exactly how those seats should be divided among each state's population. Historically, only two criteria have been generally accepted and consistently enforced in court challenges to redistricting: (1) All districts within a particular state should have roughly equal populations; and (2) Each district must be *contiguous*, meaning that every part of it must be geographically connected to every other part. The rationale for the first criterion is probably obvious: it ensures that all of the state's House members represent about the same number of people. The second criterion explains why congressional districts fit together like pieces of a jigsaw puzzle.

Beyond the puzzle-like fit of districts with equal populations, states have great freedom in deciding exactly how districts are actually drawn. To give you an idea of how much freedom, take a look at the districts for North Carolina, which are shown in Figure 15. Notice the odd shapes of some of the districts, some of which include "appendages" pushing into areas that look like they should belong to other districts. Why are the shapes so odd? You already know the answer from our discussion of Question 2 at the start of

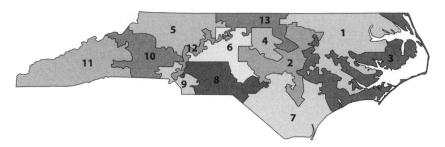

Figure 15. Boundaries for the 12 congressional districts in North Carolina as of 2010; the boundaries will be redrawn for future years based on results of the 2010 census.

this chapter: In most states, the state legislature handles redistricting, and the party and people in power do all that they can to ensure their continued grip on power.

Gerrymandering

The practice of drawing district boundaries for political advantage is so common that it has its own name—*gerrymandering*. The term originated in 1812, when Massachusetts Governor Elbridge Gerry created a district that critics ridiculed as having the shape of a salamander. A political cartoon of the time used the governor's last name to make the word "gerrymander" as a play on "salamander," and the name has stuck ever since.

Let's do a simple example to show how it works. Figure 16 shows a hypothetical state with only 16 voters (represented by little houses), half Democrats and half Republicans. The left side shows how they are distributed geographically. On first glance, it appears that the geographical distribution is even, since Democrats and Republicans live in alternating houses by row. But look what happens if you draw the districts as shown on the right. Because District 1 is 100% Republican, the Democrats end up with a clear majority in Districts 2 and 4, and a tie in District 3. If we now assume that each little house in the figure represents the general preferences of a couple hundred thousand voters, then it takes only a bit more planning for the Democrats to virtually ensure that they'll win three of the four districts, despite the 50:50 split in overall voter preferences.

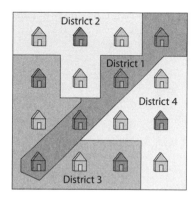

Figure 16. Gerrymandering a hypothetical state with 16 voters. (Left) The geographical distribution of the Republicans and Democrats. (Right) District boundaries that concentrate Republicans in District 1, ensuring that the Democrats will win at least two and possibly three of the four districts.

You're probably wondering: Is all this really legal? That is, just because one party controls the legislature at the time of a census, does it really have the right to divide up districts to maintain that control? The answer is both no and yes. The Supreme Court has ruled that gerrymandering for partisan political purposes violates the Constitution, which makes it illegal in principle. In practice, however, politicians can usually come up with an alternate rationale that accomplishes the same partisan effect, but to which the courts have been much less willing to object. For example, if the politicians actually said that their aim in Figure 16 was to pack Republicans into one district, the courts would not allow it. But if they said they drew the district to reflect a well-traveled business corridor, or to give a voice to a group of voters who share similar concerns about schools, then they could probably get away with it.

It's worth noting that party power is not the only objective for which gerrymandering can be used, especially when compromise is required. For example, suppose the Republicans and Democrats share power in a particular state (such as by one controlling the legislature and the other the governorship). The governor would veto a formula that heavily favors the opposing party, so they'll have to compromise—and a likely compromise will be one that gives powerful incumbents easy-to-win districts. Similarly, suppose that some powerful legislators have a grudge against some particular congressman of the other party. They might decide to redraw his district boundaries so that he no longer lives within it, making him ineligible for reelection to his current seat. Another common strategy is to draw district boundaries that force two popular incumbents from the opposing party into the same district, ensuring that they cannot both be reelected.

Complicating things even further, redistricting applies not only to congressional districts, but also to districts drawn for state legislatures. The number of possible ways of drawing district boundaries for both state and national offices is enormous. Today, the parties typically employ mathematicians or statisticians, along with computer programmers, to create sophisticated computer models with which they can test the effects of millions of different possible boundaries; they then choose the model that seems most likely to achieve their political goals without going so far as to risk losing a court challenge. If they have the power, they can put these boundaries in place; if they don't, they can make deals to accomplish as many of their goals as possible. We have indeed reached a point where, through mathematics, representatives essentially choose their own voters.

Given the degree to which redistricting is prone to political manipulation, it's natural to wonder if there is some way to stop politicians from gaming the system. Two general approaches to reform have been suggested. One would turn over the job of redistricting to independent, nonpartisan panels, such as panels of retired judges. Many other countries—including Great Britain, Australia, and Canada—use such independent panels to handle redistricting, and a handful of states do the same. Of course, no one is truly without any political bias, so this system still leaves some room for political manipulation. Perhaps a better approach would be to come up with a mathematical algorithm that would draw the boundaries independent of any human input, thereby guaranteeing that no partisan advantage could be taken. This is the option that I'd personally prefer; unfortunately, no one has yet come up with an algorithm that works well in all cases for all states. Still, either an imperfect algorithm or the independent panels would clearly be far better than our current system. We can only hope that we'll eventually find the political will to take redistricting out of partisan hands.

The Mathematics of Voting

In terms of political polarization, redistricting is without doubt the issue in which mathematics plays the greatest role. However, math turns out to have a surprisingly large role even in the basic process of voting, and I'd be remiss if I didn't at least briefly introduce you to some of the ideas.

Consider the seemingly simple process of deciding who wins an election. If there are only two candidates, the decision is easy: The one with the majority of the votes wins. But as soon as there's a third candidate, or several other candidates, everything becomes much more difficult. Take the 2000 presidential race between George W. Bush and Al Gore. This election famously came down to the vote in Florida, where the final tally showed Bush the winner by only 537 votes out of more than 6 million votes cast. The closeness of the race led to several weeks of court battles over vote-counting procedures and the need or lack of need for recounts, and it took a five-to-four decision of the Supreme Court before the winner could be officially declared. But often forgotten is that the argument was not over which candidate had a *majority* of the vote, because neither of them did. There were several other candidates on the ballot, including Ralph Nader and Pat Buchanan, and in Florida these candidates received more than 100,000 votes between them. As a result, the winner could at best have had a *plurality*, meaning the most votes but still short of a majority.

This is a crucial point, because choosing the plurality winner as the winner of the election is not the only option. In the United States, for example, many mayoral contests require a runoff between the top two candidates whenever no candidate receives a majority of the vote; runoffs are also common in national elections in many other countries. If a runoff system had been used for the 2000 presidential election, Bush and Gore would have ended up in a runoff in Florida, and the 100,000 people who had voted for other candidates would have been forced to choose between those two (or to not vote at all). There's no way to predict what would have happened, but it's very likely that the runoff would have been much less close, and all the bitterness of the court battles that followed could have been avoided.

Numerous other methods of deciding an election are also possible, and a few are in common use. For example, college football polls typically use a *point system*, in which each voter creates a list of the top twenty-five teams in rank order. A first-place ranking is worth 25 points, a second-place ranking is worth 24 points, and so on. The winner of the poll is the team with the most total points. Many clubs choose officers through what we call *sequential runoffs*, in which the candidate with the fewest votes is eliminated for a first runoff, then the remaining candidate with the fewest votes for a second runoff, and so on until only two candidates remain and one then gets a majority.

Why don't we just decide which method is best and use it for all of our elections? It's not for lack of trying. Over the past two centuries, political theorists and mathematicians have worked together to develop a set of "fairness criteria" that can be mathematically tested and that virtually everyone agrees on. Different voting methods can then be checked to see which ones meet all of the fairness criteria. The remarkable result is that a mathematical theorem, known as *Arrow's Impossibility Theorem* (after its discoverer, Kenneth Arrow), actually proves that it is impossible to develop a voting system that will satisfy all of the fairness criteria in all possible cases. In other words, we have mathematical proof that there is no clear "best" choice among the options of plurality, runoff, point system, and other ways of choosing a winner in elections with three or more candidates.

Nevertheless, some methods tend to give fair results more often than others, and plurality turns out to be one of the worst methods. You can see why with a simple and common example. Suppose that Smith, Jones, and Watson are running for governor. Let's further suppose that a large majority of voters would greatly prefer either Smith or Jones to Watson; for example, let's say that 65% of voters like both Smith and Jones and therefore would be satisfied with either of them, while only 35% like Watson. The problem is that if Smith and Jones split that 65% of the vote, Watson's 35% will represent a plurality. In other words, the plurality system makes it easy for a highly unpopular candidate to end up the winner. Different people have different opinions about what to do about this problem, but to me it seems that we'd be far better off instituting a single-runoff approach to all multi-candidate elections. The single runoff is easy to administer (which is why it is already common in mayoral races and elections in other countries), is easy to understand, and ensures that the ultimate winner is selected by a majority of voters.

Since I've come down in favor of a runoff system, I should briefly mention a variation known as *instant runoff*. The idea is that instead of having to hold a separate runoff between the top two candidates later, you ask voters to rank their preferences at the time they vote initially. Voters can rank as many candidates as they wish. For example, if you were a supporter of Jones in our above election, you'd mark a "1" next to Jones on your ballot and a "2" next to Smith, since Smith would also be acceptable to you. Watson voters would rank her 1, presumably leaving Smith and Jones unranked. Watson might still get the most first-place votes, but since she would have less than a majority, the ultimate winner would be determined by the second-place votes, and that would hand the election to either Smith or Jones. Instant runoff is actually used in a number of places around the world (and recently was rejected by voters in Great Britain), and it has many supporters. The primary drawbacks to instant runoff are that it introduces some possibilities that don't exist in a single-runoff system (particularly in elections with more than three candidates), which can in some cases allow voters to "game" the results by ranking candidates differently from their true preferences, and that it does not give the top two candidates a chance to face off against each other through additional campaign and debate time. My opinion: The only real advantage that instant runoff has over an actual runoff is that it saves the cost of holding a runoff election. Given that voting is arguably the most important thing we do in a democracy, that small savings hardly seems worth the drawbacks.

Voting Power and the Electoral System

Another important mathematical idea in voting has to do with voting power. Consider a small corporation with a total of 10,000 shares that are owned by just four shareholders. Let's call them Jack, Jill, Billy, and Lilly, and assume that Jack owns 2,650 shares, Jill owns 2,550 shares, Billy owns 2,500 shares, and Lilly owns 2,300 shares. They make all their decisions by casting votes in proportion to their shares, with each decision requiring a majority of the shares. If you look only at the numbers of shares, you might think that all four shareholders would have fairly similar power in voting, since no one has a dramatically greater number of shares than anyone else. But in fact, *Lilly has no voting power at all.* Here's why: Notice that any two of Jack, Jill, and Billy can join together to make a majority of the 10,000 shares, so it takes only two of them to make a decision. Lilly, however, cannot form a majority with any one of them; for example, if she joins forces with Jack (who has the most shares), they can still be outvoted by the combination of Jill and Billy. Therefore, Lilly can only be part of a majority if at least two of the others join with her, and because those two would have had the majority even without her, her vote is completely superfluous.

Real-life cases of voting power are rarely quite so extreme, but there are many cases in which the power distribution can be very surprising. Suppose, for example, that some particular bill is supported by 49 senators and opposed by 49 others, with two not having a strong opinion. In that case, the two essentially hold all the power, because neither side can get a majority without them. You can probably imagine the ways in which they might barter their votes. This type of situation is particularly common in countries with parliamentary systems. In Israel, for example, small religious parties have exerted clout far beyond what their numbers would suggest because the larger parties often need their votes to form majorities on major issues.

The issue of voting power also comes into play in U.S. presidential elections. As you are probably aware, the presidency is decided not by the popular vote but rather by the *electoral* vote. As mandated by the Constitution, each state gets as many electoral votes as it has members of Congress (senators *plus* representatives). These facts skew the voting power in at least two ways.

First, because every state has two senators and at least one representative, small states have more voting power per person than larger states. In 2008, for example, California had 55 electoral votes for its population of nearly 37 million people, while Wyoming had 3 electoral votes for its population of a little over 500,000 people. If you do the division, you'll find that each electoral vote in California represented almost four times as many people as each electoral vote in Wyoming, which means Wyoming voters had much greater voting power per person.

A second way in which the electoral system skews voting power generally favors large states, and it arises from the fact that nearly all states (the exceptions are Maine and Nebraska) use a winner-take-all method for distributing their electoral votes. The 2000 election offers a perfect example. Because the rest of the national electoral vote was quite close, Florida's relatively large cache of electoral votes made it the deciding factor in the election. As a result, Bush's 537-vote margin of victory in Florida ended up counting more than Gore's half-million-vote advantage in the nationwide popular vote.

Does this skewed voting power make the electoral system a bad idea? Not necessarily. There are some good reasons that it was put in the Constitution in the first place, and it would take a constitutional amendment to completely eliminate it. However, for those who would prefer a system in which the president must win the popular rather than the electoral vote (and I count myself among this group), a few clever people have found a mathematical trick that would accomplish it without the need for a constitutional amendment. Imagine that states holding a majority of the electoral votes among them decided that each one would give its electoral votes to the winner of the national popular vote. Then whoever won the national popular vote would automatically end up with enough electoral votes to win the election. Several states have already passed or are considering legislation that would do just this *if* enough other states did the same. Time will tell, but I suspect that if we have more elections in which the electoral and popular vote winners are not the same, we'll end up seeing the electoral system effectively overridden by this recent mathematical idea.

Math for Life

The fact that so much mathematics is involved in voting and the politics of redistricting comes as a great surprise to most people, but it is just another illustration of the important mathematical ideas that you probably didn't learn in school. Even in those rare high school and college mathematics courses that cover these issues at all, I've found that most instructors focus on systems for choosing an election winner and on apportionment of representatives, leaving out the issue of redistricting. The problem with this approach is that while all of these issues make for interesting mathematical discussions, redistricting is the one that has by far the greatest implications for our lives. Because it is left out of the curriculum, very few people understand its role in our increasingly polarized politics, or how easily it is manipulated for advantage by those in power.

In closing this chapter, I urge you to pay much more attention to the role of mathematics in politics. Watch for elections in which a plurality rule leads to an unpopular winner. Look for cases where voting power is not fairly distributed. Most importantly, pay attention to the way that redistricting is done, particularly if you are reading this book shortly after its publication, when most states will be deep in the redistricting process. The extremists have gained the upper hand by learning to manipulate the mathematics of redistricting. Only by understanding what they do can we restore sanity to our political process.

9

The Mathematics
of Growth

Somehow we just missed that home prices don't go up forever.
— **Jamie Dimon, CEO of J.P. Morgan Chase**

The greatest shortcoming of the human race is its inability
to understand the exponential function.
— **Albert A. Bartlett**

We'll start this chapter a little differently, with a short parable and a series of questions based upon it. The parable and questions were originally developed by physics professor Albert A. Bartlett of the University of Colorado, also the source of our second quote above. I have modified only the details of the storytelling.

The Parable of the Bacteria in a Bottle: Once upon a time, at precisely 11:00 p.m., a single bacterium was placed into a nutrient-filled bottle in a laboratory. The bacterium immediately began gobbling up nutrients, and after just one minute—making the time 11:01—it had grown so much that it divided into two bacteria. These two ate until, one minute later, they each divided into two bacteria, so that there were a total of four bacteria in the bottle at 11:02. The four bacteria grew and divided into a total of eight bacteria at 11:03, sixteen bacteria at 11:04, and so on. All seemed fine, and the bacteria kept on eating happily and doubling their number every minute, until the "midnight catastrophe." The catastrophe was this: At the stroke of 12:00 midnight, the bottle became completely full of bacteria, with no nutrients remaining—which meant that every single one of the bacteria was suddenly doomed to death.

We now turn to our questions as we seek to draw lessons from the tragic demise of the bacterial colony.

Question 1: The catastrophe occurred because the bottle became completely full at 12:00 midnight. When was the bottle *half* full?

Answer: If you pose this question to most any high school or college class, the vast majority of students will answer, "11:30." After all, we know the bottle was empty at 11:00 and full at 12:00, so it seems reasonable to guess that it was half full after half the time. But this is not the case. Because the bacteria double in number every minute, they must also have doubled during the last minute. In other words, the bottle must have been half full just one minute before the end, which means it was half full at 11:59.

Question 2: You are a mathematically sophisticated bacterium, and at 11:56 you recognize the impending disaster. You immediately jump on your soapbox and warn that unless your fellow bacteria slow their growth dramatically, the end is just four minutes away. Will anyone believe you?

Answer: Note that our question is *not* whether you are correct, because you clearly are—we already know that the end really does come at 12:00. Rather, we're asking if anyone will *believe* you, and to address that question we must think of how the situation will appear to your less mathematically sophisticated friends. As we've already seen, the bottle would be half full at 11:59. Continuing to work backward through the doublings each minute, we find that it would be one-quarter full at 11:58, one-eighth full at 11:57, and one-sixteenth full at 11:56. Therefore, if your fellow bacteria look around the bottle at 11:56, they'll see that only one-sixteenth of the bottle's space has been used. That means a 15-to-1 ratio of unused space to used space, so you are in essence asking them to believe that, in just the next 4 minutes, they'll use up 15 times as much space as they did in their entire 56-minute history so far. Unless they do the mathematics for themselves, they are unlikely to take your warnings seriously. Fig-

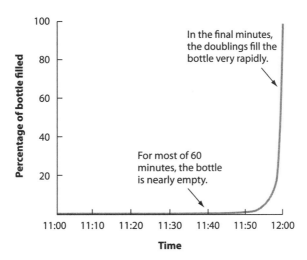

In the final minutes, the doublings fill the bottle very rapidly.

For most of 60 minutes, the bottle is nearly empty.

Figure 17. This graph shows how the percentage of the bottle that is full changes during the one hour between the first bacterium and the catastrophic end of the bacterial colony.

ure 17 shows the situation graphically. Note that the bottle remains nearly empty for most of the 60 minutes, but the continued doublings fill it rapidly in the final 4 minutes.

Question 3: It's 11:59 and, with the bottle now half full, your fellow bacteria are finally taking your warnings seriously. Hoping to avert their impending doom, they quickly start a space program, sending little bacterial spaceships out into the lab in search of new bottles. To their relief, they discover that the lab has three more bottles that are filled with nutrients but have no one living in them. They immediately commence a mass migration through which they successfully redistribute the population evenly among the four bottles (the three new ones plus the one already occupied), just in time to prevent the midnight catastrophe. How much more time do the additional bottles buy for their civilization?

Answer: Even after discussing the first two questions, I've found that most students still assume that the three new bottles buy the bacteria three more hours. Their reasoning is simple: They assume that because it takes one hour to fill one bottle, three more bottles should mean three more hours. But remember that the bacterial population is still *doubling* each minute. Therefore, if there are enough total bacteria to fill one bottle at 12:00, there will be enough to fill two bottles by 12:01, and four bottles by 12:02. The discovery of three new bottles buys them *only two additional minutes.*

Question 4: Because the three extra bottles bought so little time, the bacteria keep searching out more and more bottles. Is there any hope that additional discoveries will allow the colony to continue its rapid growth?

Answer: It's tempting to assume that if they work hard enough, they can keep making new discoveries that will allow them to survive; after all, it's nice to think that "nothing is impossible." But let's do some simple calculations. The colony started with 1 bacterium, so when it divided at 11:01, there were $2^1 = 2$ bacteria. As the growth and divisions continued, there were $2^2 = 2 \times 2 = 4$ bacteria at 11:02, $2^3 = 2 \times 2 \times 2 = 8$ bacteria at 11:03, and so on. Following the pattern, there must have been 2^{60} bacteria after 60 minutes (when the bottle was full). You can use a calculator to figure out that 2^{60} is 1.15 million trillion (just push the buttons for 2^60), but what we really care about is how much space the bacteria are taking up. Let's assume that each bacterium looks approximately like a microscopic cube that is one ten-millionth of a meter, or 10^{-7} meter, on a side. Volume is length times width times depth (and for a cube all three are the same), so we find that each bacterium has a total volume of 10^{-21} cubic meters. Multiplying this volume per bacterium by the 2^{60} bacteria that are in the bottle, you'll find that their total volume after one hour was indeed about that of a small bottle.

Now, suppose that the bacteria are somehow able to keep their doublings going until 1:00 a.m. At that time, it has been 120 minutes since the colony began, so the number of bacteria is 2^{120}. If you multiply this number by the volume of each bacterium, you'll find that after this second hour of growth, the bacteria occupy a volume so large that they cover *the entire surface of the Earth* in a layer more than 2 meters (6 feet) deep.[24]

Taking the idea further, the same simple calculations show that after a mere five and a half hours of these doublings, *the calculated volume of the bacteria would exceed the volume of the known universe.* Clearly, this cannot happen. Aside from the fact that there's nothing for the bacteria to eat in space, Einstein's theory of relativity tells us that it is not possible to travel faster than the speed of light, so there would be no way for the bacteria to travel across the universe in those mere five and a half hours. In other words, much as we like to say that "nothing is impossible," some things are—and indefinite growth through repeated doublings is one of them.

Doubling Times

You might think that growth through successive doublings would be a rare case, but it is actually quite common. Growth marked by repeated doublings is called *exponential* growth,[25] and it occurs any time something grows by a fixed percentage per unit time. The best way to understand it is by contrasting exponential growth with *linear* growth, which is growth by a fixed amount (rather than percentage) per unit time.

Imagine two communities that each start with the same population of 10,000 people. Lintown grows linearly by 500 people each year, so its population reaches 10,500 after one year, 11,000 after two years, 11,500 after three years, and so on. Now consider Expotown, which also starts with 10,000

---

24. If you want to see this for yourself: Multiplying 2^{120} bacteria by the volume of 10^{-21} cubic meters per bacterium gives a total volume of approximately 1.3×10^{15} cubic meters. Now, divide this volume by the total surface area of the Earth, which is about 5×10^{14} square meters. The result is about 2.5 meters, which means that is the depth to which the bacteria would cover the entire Earth.

25. To understand the term, recall that when we write powers, such as 2^{60}, the power (60, in this case) is often called the *exponent*. As the parable of the bacteria shows, the exponent increases with each doubling, which is why we call it *exponential* growth.

people but grows exponentially by 5% each year. Because 5% of 10,000 is 500, it ends the first year with the same population of 10,500 as Lintown. In the second year, however, Expotown's population increases by 5% of 10,500, which is 525, so it ends the year with a slightly larger population of 11,025. In the third year, Expotown's population grows by 5% of 11,025, or 551 people, to 11,576. Notice that the fixed *percentage* rate of growth means that Expotown's numerical growth becomes larger with each passing year.

It's easy to continue calculating the annual populations for each of the two towns, and Figure 18 shows the results for more than 40 years' worth of growth. You should focus on two key facts. First, Expotown's growth outpaces Lintown's more and more as time goes by; for example, after 42 years Expotown's population will be nearly 80,000, while Lintown's will be only 31,000. Second, notice that Expotown's total population is doubling approximately every 14 years. It reaches 20,000 after about 14 years, 40,000 after about 2 × 14 = 28 years, and 80,000 after 3 × 14 = 42 years. This is the key insight we were looking for: The fact that Expotown is growing by the same 5% each year means that its population grows with a characteristic *doubling time*—about 14 years in this case.

You will find repeated doublings with any growth by a fixed percentage, including compound interest on a bank account, resource use that is growing

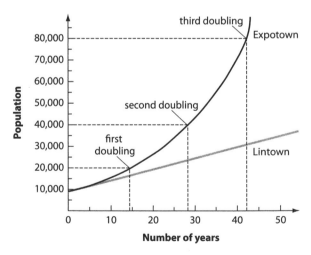

Figure 18. Lintown and Expotown both start with 10,000 people, but Lintown grows linearly by 500 people per year, while Expotown grows exponentially by 5% per year. Notice that Expotown undergoes a doubling of its population approximately every 14 years.

by some percentage each year, and many other quantities. The only thing that varies is the amount of time required for each doubling. The higher the percentage rate of growth, the faster the doubling time. With growth of 5% per year, we've found a doubling time of about 14 years. With growth of 100% per year (like the bacteria if they divided each year instead of each minute), the doubling time would be one year.

Although it's mathematically equivalent to work with percentage rates of growth or doubling times, it's often easier to visualize growth by focusing on the doubling time. For example, suppose you want to predict Expotown's population after 140 years, assuming no change to its growth rate. Mentally, it's a bit tricky to calculate what you'll get with 5% annual growth for 140 years, although there's a simple algebraic formula that you can use.[26] But given that the doubling time is about 14 years, you can immediately see that 140 years is approximately ten doubling times, which means the population will have increased by a factor of close to $2^{10} = 1,024$. In other words, the population will have become about 1,000 times the starting population of 10,000, which means that in 140 years, Expotown will have become one of the world's largest cities, with a population of close to 10 million people. (Meanwhile, Lintown would have grown only to 80,000 people.)

For any case in which the percentage rate of growth is relatively low, there's a wonderful little formula for finding the approximate doubling time. The formula is called the "rule of 70" because it says that *the doubling time is approximately 70 divided by the percentage growth rate.* For example, the rule of 70 tells us that the doubling time for Expotown, with its 5% annual growth rate, should be approximately $70 \div 5 = 14$ years—just as we've already found it to be. Be careful, though: The rule of 70 works well when the growth rate is about 5% or less, but less well as the growth rate gets higher. For rates above about 15%, it breaks down almost completely. In those cases, there is a more complex formula you can use to find the exact doubling time, but it's usually easier to use the algebraic formula for exponential growth (see note 26).

---

26. Here's the formula: Amount after time t = (starting amount) $\times (1 + r)^t$, where r is the fractional rate of growth. For the Expotown case, the rate of growth is $r = 5\% = 0.05$ per year, the starting amount is the initial population of 10,000, and the time is $t = 140$ years. Therefore, the population after 140 years is $10,000 \times (1 + 0.05)^{140}$, which is about 9.3 million.

"Somehow we just missed that . . ."

It's time now to return to the housing bubble, which we discussed way back in Chapter 1. Real estate investment boomed during the bubble because people thought they could count on a high rate of return. The precise rate at which home prices rose during that period varied both by year and geographically, but throughout the bubble it averaged close to 15% per year, which translates to a doubling time of about 5 years. Let's see what this really means.

According to the Census Bureau, the median home price in 2005 (for the United States) was close to $250,000. Therefore, if the bubble had not popped and the doubling time had remained about 5 years, the median price would have reached $500,000 by 2010. Continuing into the future, the median price of a home would have become $1 million by 2015, $2 million by 2020, $4 million by 2025, and moving out a bit further, $128 million by 2050. Could anyone really think that half of today's children will be able to afford houses costing more than $100 million when they reach middle age? For that matter, as we discussed in Chapter 1, there's no way that median housing prices could even have risen to $500,000 in 2010, because the average family does not have enough income to justify that kind of average price.

In essence, the growth of home prices during the bubble was no different from the growth of the bacteria in the bottle. It was fun while it lasted, but it had to stop. *Exponential growth always comes to a stop.* It's not because there's anything wrong with it, or because it becomes unpopular, or anything else. Rather, exponential growth always stops because it is physically (or economically) impossible for it to continue indefinitely.

How, then, could so many people have believed that housing prices would continue an inexorable rise? The answer is essentially the same one we found to Question 2 for the bacteria in the bottle: The last moments of exponential growth come at you so fast that, unless you do the calculations, you probably won't see them coming. Remember, by doubling in its final minute, the bacterial colony filled as much new space as it had used in its entire first 59 minutes. In much the same way, housing prices during the last years of the bubble grew almost as much as they had in the previous several decades.

Let's briefly pause to summarize the two most important lessons we've learned about exponential growth:

1. Exponential growth cannot continue forever. It can only be a temporary phase in the growth of any real quantity, whether it is a population, home prices, the value of some other investment, or anything else.
2. Exponential growth has a way of creeping up on you, because each successive doubling represents approximately as much total increase in number (or value) as all previous increases combined.

Now we come to the heading of this subsection, which is taken from the Jamie Dimon quote that appears at the beginning of this chapter. I've never met Mr. Dimon, and I feel a bit bad about singling him out, but . . .

While it's true that exponential growth creeps up on you, all it takes to see it coming is the simple arithmetic that we've been doing in this chapter. Again, the reason your fellow bacteria had a hard time believing you when you stood on your soapbox at 11:56 wasn't that you were wrong, it was that they didn't do the math. If they had, they would have seen as clearly as you that a catastrophe was imminent unless action was taken. In much the same way, there was no room for any doubt that housing prices had to stop their rapid rise. Perhaps we can excuse the fact that so much of the public missed this point, since our schools do such a poor job of teaching about exponential growth. But Jamie Dimon was the *CEO of one of the largest financial corporations in the world*. I don't know which is worse: the possibility that he might not have been truthful in sworn testimony, or the scary thought that someone in his position couldn't do math at the level we've done it in this chapter.[27]

World Population Growth

Let's next take a look at the way human population has grown with time. Figure 19 shows estimated world human population over the past 12,000 years. You can find many books and articles that will give you the details of how and why the population has changed with time, but here we'll focus on one key

---

27. To be fair, Mr. Dimon's quote does not explicitly refer to home prices going up *exponentially* forever. However, this was almost certainly implied, since "going up" in economic terms generally means at a rate faster than inflation, which means by some percentage every year.

World Population, 10,000 BCE to Present

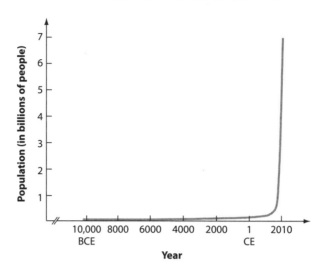

Figure 19. Estimated world human population over the past 12,000 years.

fact: If you look back at Figure 17, you'll see that aside from the scales on the axes, the human population graph looks almost identical to the graph for the bacteria in the bottle.

It's easy to understand why human population tends to grow exponentially. Without interventions such as birth control, we have a tendency to produce more children than there are parents. If these children survive to adulthood and have their own children, then our numbers must grow with each generation. For example, suppose that, on average, every couple has three children that survive to adulthood. In that case, the population would rise 50% with each successive generation, and as we've discussed, anything that grows by a percentage rate is growing exponentially.

In discussing population growth, it's usually easier to focus on the annual growth rate than the generational growth rate. To do this, we simply subtract the death rate from the birth rate. Globally averaged, the current annual birth rate (as of 2011) is about 19 per 1,000 people, or 19/1,000; that is, if you pick 1,000 people at random from around the world, on average they'll give birth to 19 babies in the next year. The current annual death rate is about 8 per 1,000 people, or 8/1,000, which means that on average 8 of your randomly selected 1,000 people will die in the next year. The global growth rate is therefore $19/1,000 - 8/1,000 = 11/1,000$, which is the same as 0.011 or 1.1%. Using the rule of 70, this translates to a doubling time of about 64 years.

Let's see what this means. To make the arithmetic a little easier, let's call the doubling time about 60 years and start from the world population of close to 7 billion at the end of 2010. If the annual population growth rate held steady at 1.1%, then the human population would double to 14 billion by 2070, double again to 28 billion by 2130, and reach 56 billion by 2190.

It's probably already obvious that our numbers cannot really go so high, but let's keep going anyway, just to see where it leads. The total land surface area of Earth is about 150 million square kilometers. If you extend human population growth with a growth rate of 1.1%, you'll find that, in a mere 900 years, our entire planet would be standing room only. Even if we could somehow colonize the rest of the planets and moons in our solar system, we'd have standing room only on all of them, too, just a couple hundred years after that. And like the bacteria in the bottle, we'd reach the impossible limit of filling the universe within a few thousand years.

People can and do argue about the relative merits of population growth versus population stability, or about how many people Earth can actually support. But these arguments are a luxury of our time. After many thousands of years of human civilization, our numbers have reached the point where they cannot continue their exponential growth much longer. We cannot really stand elbow to elbow on every bit of land on Earth, so a roughly 1% growth rate cannot possibly continue for another 900 years. It's almost inconceivable to imagine the planet supporting a population of 56 billion, so this type of growth for another 180 years is also out of the question. In fact, very few experts believe that our planet could sustain a population of 14 billion people, which means that our current growth cannot continue even for another 60 years.

The lesson should be clear. We are like the bacteria in the bottle a minute or two before midnight, because one way or another, the exponential growth of world population will stop within the next few decades. Fortunately, we differ from the bacteria in one crucial way: While there is nothing that either we or the bacteria can do to change the fact that exponential growth will stop, we have the ability to choose *how* it stops. There are two basic ways to slow or stop the growth of a population: (1) decrease the birth rate, or (2) increase the death rate. Because no one wants to see an increase in the death rate, our only real choice is to reduce the birth rate.

The good news is that most people are already choosing to reduce the birth rate. Before the twentieth century, the average woman gave birth to more than 6 children during her lifetime. By 1950, that average had fallen to about 5 children. It has fallen more rapidly since then, and today the global average is about 2.5 children. It needs to fall just a little more (to about 2.0 to 2.1 children, depending on assumptions for death rates) for world population to stop growing and stabilize; in some nations, including Japan and several nations in Europe, the birth rate has already fallen below this threshold, leading to declining populations. If the downward trend in birth rates continues to follow its pattern of recent decades, human population will stabilize at about 9 to 10 billion people by around 2050.

Before we become too complacent, however, we should keep two important facts in mind. First, even if the population does indeed stabilize at 9 to 10 billion by 2050, that still means an increase of 2 to 3 billion people over the next 40 years. That's an average of 50 to 75 million more people each year. At 75 million, which is about the current annual increase (it would begin to drop as the population leveled out), we find that *every four years the world is adding a population nearly equivalent to that of the entire United States.* Think about the infrastructure, food, education, energy, and jobs that will be needed for all these people, and you can see that we face a daunting global challenge.

This brings us to the second important fact, which is that this daunting scenario is probably the *best*-case scenario, and it will happen only if people all around the world make a conscious decision to make it happen. After all, having fewer children is a choice that requires action such as the use of birth control; without such action, the birth rate would go back to its natural high average of 6 or more children per woman. So the next time you get involved in a debate about family planning, birth control, teen pregnancies, single motherhood, or any other population-related topic,[28] please remind everyone of the following indisputable facts about world population growth.

---

28. Please note that none of this need have any linkage to the much more emotionally charged issue of abortion, because well-constructed family planning efforts should always promote planning that avoids unwanted pregnancies in the first place.

The exponential growth of world population will come to a stop, and it will come to a stop soon. It will do so through either a decrease in the birth rate or an increase in the death rate. This statement is not a threat, a warning, or a prophecy of doom. It is simply a law of nature, because exponential growth always stops. As human beings, we can choose to bring it to a gradual halt through conscious effort to reduce the birth rate to the point of population stability. But if we do not make this conscious choice, then the death rate will have to go up dramatically, and this is a fact over which we have no more control than we do over hurricanes, tornadoes, earthquakes, or the explosions of distant stars. In the end, then, we need only answer the simple question of whether we can be smarter than the bacteria in the bottle.

Live Long and Prosper

The ideas we've covered here also apply to our economy at large. We generally measure economic growth as a percentage per year, which means that our economy has also been growing exponentially. This should not be surprising. If our standard of living remained perfectly unchanged, our economy would still grow at the same rate as the population (which has been growing exponentially in the United States at a rate somewhat below the rate for the world as a whole). Because our standard of living has risen with time, the actual rate of economic growth has been greater than the rate of population growth.

The problem we face is that, at least the way we measure it today, most of our economic growth depends on the use of physical resources. Because physical resources are limited, the exponential growth in our use of them cannot continue indefinitely. We therefore conclude that not only world population growth but also resource-based economic growth must soon stop.

This simple fact is not as pessimistic as it sounds or as it is often made out to be. For one thing, although the exponential growth of population and resource usage must both stop "soon," they need not necessarily stop at the same time. No one knows exactly how far our resources can stretch, but I'm personally optimistic that there are enough so that we can substantially raise the standard of living for all of the 9 to 10 billion people who will live on this planet in a few decades. If we can stop our population growth there—or better yet, even before we reach that number—then by the turn of the next century we might have a world that is far healthier, happier, and more peaceful than the one we have today.

For another thing, I've used the words "resource-based economic growth" for a reason, namely, that not all "growth" requires increased use of resources. In fact, as we discussed for energy in Chapter 7, resource-based economic growth often has consequences that can make us *less* rather than more well off. A better measure of our well-being would look at some combination of health and happiness, which we can consider together as *prosperity*. Prosperity and resource-based growth are not the same thing, as we can see simply by looking at our own individual lives. While we all grow physically during childhood, once we become adults we don't want to grow physically any longer, since any such growth is usually in the wrong direction. However, we do still want to prosper through health and happiness.

The bottom line is that laws of mathematics leave us no choice. We cannot continue to have resource-based economic growth forever, so if we use that as a measure of our well-being we will ultimately be disappointed. Therefore, we need to redefine how we measure prosperity. After all, our goal should be to achieve what Mr. Spock of *Star Trek* always suggested: Live long and prosper.

Exponential Decay, Nuclear Waste, and the Age of the Earth

We've been focusing on cases of exponential *growth*, but it is also possible for quantities to decrease in number, or decay, exponentially. Just as exponential growth occurs with anything that grows by a fixed percentage per unit of time, exponential decay occurs when something declines by a fixed percentage. In that case, instead of finding a characteristic doubling time, we find a characteristic time for the quantity to decrease to half of its prior value, and we call this time the *half-life*. For example, suppose that the population of an endangered species is declining by 5% per year. Decay is equivalent to a negative growth rate, and you'll recall that the rule of 70 gave us a doubling time of about 14 years for a population growing at 5% per year. The same rule tells us that a population that is declining by 5% per year will have a half-life of about 14 years. If the endangered population is 10,000 today, it will fall to 5,000 in 14 years, to 2,500 in 28 years, to 1,250 in 42 years, and so on.

Although they are not always as obvious as cases of exponential growth, we can find plenty of cases of exponential decay. The concentration of a drug in your bloodstream tends to decay exponentially; for example, if a medicine has a half-life in your body of 12 hours, then you can maintain a reasonably steady concentration of it by taking it twice a day. Declining populations, such as those of endangered species, often decay exponentially, and the remaining amount of some limited resource, such as oil, may also follow an exponential decay pattern.

While the above examples indicate *approximately* what we expect from exponential decay, there is one thing that follows the mathematical pattern of exponential decay with near-perfect precision: radioactive material. Every different radioactive substance decays at a precise rate that can be measured in laboratories. Once the decay rate is measured, which can be done in a few months for even the longest-lived radioactive materials, we can translate it into a half-life. That is why you usually hear radioactive materials described by their half-lives. For example, in the release of radiation from Japanese nuclear reactors after the March 2011 earthquake, one substance of particular concern was radioactive iodine-131, which has a half-life of 8 days. This half-life tells us that half of the released iodine-131 would have decayed after 8 days, which means that only the other half would remain radioactive and dangerous. The amount remaining dangerous would be 1/4 of the original amount after 16 days, 1/8 after 24 days, 1/16 after 32 days, and so on. By a few weeks after the release, so little iodine-131 would remain that there would no longer be any danger from it.

Unfortunately, many radioactive materials can pose dangers for much longer. Another concern was cesium-137, which has a half-life of 30 years and therefore remains dangerous for many decades. The region around the Chernobyl nuclear power plant, where a major release of radioactive material occurred in 1986, still has dangerous levels of cesium-137 (as well as of numerous other materials).

In Chapter 7, I mentioned the difficulty of disposing of the nuclear waste produced as a byproduct of successful operation of nuclear power plants. To see why it poses such a problem, consider just one major ingredient of nuclear waste: plutonium-239, which has a half-life of about 24,000 years. With such a long half-life, any storage solution for this waste product needs to keep it away from the population and environment for at least a few hundred thousand years before it will have decayed to the point where it is no longer dangerous.

There's also a more positive side to radioactive materials and half-lives: They allow us to determine the ages of ancient artifacts, fossils, and rocks. To see how this works, consider the radioactive form of carbon that is called carbon-14. Carbon-14 is continually and naturally produced in the atmosphere (in small enough amounts that it is not considered dangerous), and it is then incorporated into living organisms through their respiration. When an organism dies, it no longer takes in any new carbon-14, and the carbon-14 it contains decays with a half-life of about 5,700 years. Now, suppose you find an old bone, or tree, or clay painted with organic dye, and you can determine that half of the carbon-14 it contained at the time of death has decayed.[29] Then, voilà, you can conclude that it died 5,700 years ago. Similarly, if you find an artifact that contains only 1/8 of the carbon-14 that it contained originally, then you know it was made three half-lives of carbon-14 ago (since it takes three half-lives for the value to fall to 1/8 of the original amount), which means $3 \times 5,700 = 17,100$ years ago.

We can be very sure that this technique of *radiometric dating* is reliable, because we can verify it through independent techniques. Carbon-14, for example, has been used to date numerous Egyptian artifacts that have actual dates printed on them, and the dates found with carbon-14 match the printed dates. For somewhat older artifacts, we can sometimes use tree rings to get independent dates; again, the ages found with the tree rings and the carbon-14 dating agree.

The relatively short half-life of carbon-14 means we can't use it to date anything older than a few tens of thousands of years. But many other naturally occurring radioactive materials have much longer half-lives. For example, uranium-238, which decays into lead, has a half-life of 4.47 billion years. We

---

29. In case you are wondering how we can know the amount of a radioactive substance that was originally present: For carbon-14, we can measure the current rate at which it is produced in the atmosphere (which determines how much is incorporated by respiration into organic matter today), and we can estimate its past production rates by studying such things as tree rings. We can be even more precise for many other radioactive materials, because they are often found intermixed with their characteristic decay products, making it possible to calculate the exact starting amounts of the radioactive materials.

have found moon rocks that contain about half of their original uranium-238, while the other half has become lead; that is how we know that the Moon is close to 4½ billion years old. Again, there are independent checks that can be made, such as using other radioactive substances found in the same rock to confirm that the ages found with the different materials agree.

This brings up an important issue. The techniques we use to determine the ages of rocks and fossils are mathematically equivalent to the techniques used to calculate compound interest on your bank account. No one doubts interest calculations, yet polls show that a substantial fraction of the public does not believe that Earth is really as old as radiometric dating shows it to be. Perhaps even more important, the theory of evolution is strongly supported by the fact that we see gradual change in the fossil record, and this record is dated by radiometric techniques. Thus, while things like the age of the Earth and the evidence for evolution often come up in the context of science and religion, they are largely mathematical issues. Unless you're willing to ignore the simple arithmetic that we've discussed in this chapter, there's no escaping the fact that rocks and fossils support the idea that species have evolved over time on a 4½-billion-year-old Earth.

Math for Life

If you look back at the second quote at the beginning of this chapter, you can probably now see why Professor Bartlett considers an understanding of the exponential function—by which he means the mathematics of exponential growth—to be so important. In just these few short pages, we've seen how the exponential function explains the popping of the housing bubble, the fact that world population growth must soon come to a stop, and the dangers of radioactive waste, and its applications go far beyond those we've had time to discuss.

I'll quibble with just one small aspect of Professor Bartlett's quotation: I think he misspoke in citing the "inability" of the human race to understand exponential growth, because he himself has done such a superb job of explaining it to tens of thousands of people; in fact, he's given his famous lecture on "arithmetic, population, and energy" more than 1,600 times. From my own standpoint, I hope this chapter has shown you that while there's no doubt that exponential growth has surprising properties, it's not at all difficult to understand.

As usual, this brings us to the question of why, given its importance and simplicity, so few people currently understand the exponential function. As we've seen, even well-educated people (Mr. Dimon being our case in point) can "somehow miss" its obvious implications. Moreover, unlike numerous other topics we've covered in this book, the exponential function is a standard part of the algebra curriculum, which means most people spend at least some time on it in high school.

My own answer to the question is that even though exponential growth is covered in the curriculum, it's covered in such a way that the most important points get buried. In particular, algebra students are taught the mathematical *content* of exponential growth, but they often miss out on the far more important *context*. I attribute this problem to the difference between what I call *content-driven* and *context-driven* approaches to the teaching of mathematics.

The current mathematics curriculum is almost universally taught using a content-driven approach, meaning that teachers begin with the purely mathematical content. For example, most teachers begin the study of the exponential function by teaching students how to work with it in its most general form. Only later do they discuss the applications, such as compound interest and population growth, that provide the context and the lessons we've discussed. The problem with this approach is that because students are not given the context for what they are learning in advance, many of them get lost in the algebra before ever reaching the more important parts.

For that reason, I believe we'd be far better off if we turned things around, using a context-driven approach in which we start with the things that students will find most meaningful. In this book, for example, we talked about compound interest way back in Chapter 4, before I even told you that it is an example of the exponential function at work. Then, when we began our more direct discussion of exponential growth in this chapter, we started with the easy-to-understand but crucial lessons from the bacteria in the bottle, and went on to apply those lessons to more practical examples. We've stopped at this point, but if this were a class in algebra or quantitative reasoning, you'd now be ready to explore the more general mathematics of the exponential function. In other words, the context-driven approach can allow us to cover just as much material as the traditional content-driven approach, but because we start with the context, we make it much easier for students to succeed in learning it.

Now, before you start thinking that I've come up with some remarkable insight about teaching, I should point out that I'm not the first to suggest this idea. In fact, the context-driven approach is really just a reflection of the well-known fact that we learn better by starting with concrete ideas and moving to abstract ones than the other way around. You can see this approach even in the writings of Plato and other ancient Greek philosophers, so it's clearly not a new insight. In the mid-twentieth century, the famed psychologist Jean Piaget did experiments demonstrating that children can learn concrete ideas years before they are ready to learn abstractions, and other research shows that adult brains still learn best by going from concrete to abstract.

The basic problem, then, is that our schools have somehow managed to create curricula that present mathematics in a way that is directly contrary to what thousands of years of experience show works best. It's actually not that hard to see how this came about, but that's not a topic for this book. Rather, the point here is that we know there's a better way, and the better way isn't even hard to institute. We just need to reform our curricula, and rewrite our textbooks, so that they move from concrete context to abstract content, rather than the other way around.

Epilogue

Getting "Good at Math"

Human history becomes more and more a race between
education and catastrophe.
— H. G. Wells

Facts do not cease to exist because they are ignored.
— Aldous Huxley

Question: If we want to make more people "good at math," the process clearly starts in elementary school. The best way to improve the teaching of math at the elementary level is:

Answer choices:
a. Use a "constructivist" curriculum that emphasizes conceptual understanding of math, rather than memorization of facts
b. Stop wasting time on constructivist nonsense and make sure kids memorize their addition and multiplication facts
c. Test students every year so that we'll know how they are faring in their mathematics work
d. Provide teachers with more time for class preparation and participation in teaching workshops
e. Recruit and retain outstanding teachers who can get students to spend more time studying

Unlike the questions with which we started previous chapters, this one does not have an unambiguous mathematical answer; it is essentially a matter of opinion. Nevertheless, I'd argue that one answer makes far more sense than all the others. Hint: It's the only one that, as a nation, we really haven't yet tried.

Let's start with A and B, which have been the topic of the so-called math wars that have erupted in numerous communities across the nation. In fact, if you type "math wars" (with the quotes) into a Web search box, you'll turn up dozens of articles and discussions about whether it is better to spend time memorizing facts or "constructing" an understanding of those facts. It makes for some entertaining reading, but I find the entire debate to be completely ridiculous. It's analogous to arguing about whether it's better to teach kids how to read individual words on a page or to understand what the words mean—it seems obvious that success in reading requires both. The same is true for mathematics, and it doesn't take long to see why. In my own kids' elementary school, the focus has been on constructivism, and it has resulted in numerous kids graduating elementary school without knowing their multiplication facts. Perhaps thanks to the focus on understanding, some of these kids become quite good at figuring out what they need to do to solve a mathematical problem. Unfortunately, their lack of basic skills means they then become stuck on the simple calculations needed to carry out the solution, which leaves them immensely frustrated and therefore prone to grow up believing that they are "bad at math." On the flip side, you'll never be able to apply mathematics in the ways we've applied it in this book if you've only memorized the facts without actually understanding what they mean. So neither A nor B is the answer we seek, since math teaching should be a balanced combination of both.

Option C has now been implemented almost everywhere, since nearly all public school students are now required to take annual tests to measure their progress. I'm a big believer in tests, though I worry that the current implementation is causing almost as many problems as it solves. For example, if you really want to make sure students are progressing, you should give them tests in class on a regular basis, such as every one to two weeks, but I've heard of many cases in which the demands of the standardized tests have caused teachers to give fewer "regular" tests. I've also seen many sample questions from the standardized tests that are confusing, ambiguous, or just poorly written—and low-quality tests can only provide low-quality data. Then there is the well-known problem of teachers "teaching to the tests" while neglecting anything that does not appear on the tests—which means neglecting many of the topics that many students find most inspiring. In any event, whether or not you agree with current testing policies, it's clear that an annual test cannot be the solution to a teaching problem; at best, it can only help us identify the problems to be solved.

That leaves us with options D and E, which are both about teachers. Option D essentially suggests helping existing teachers to get better. This is clearly a good idea (as long as it does not come at the expense of class time for students), but it is not a panacea. The simple fact is that some people are better teachers than others. We can argue about why—for example, are they smarter, or more dedicated, or more naturally gifted?—but everyone knows that some teachers are phenomenal while others are not, and I doubt that any amount of workshop and prep time can turn all the "nots" into phenoms.

So if we really want to improve the teaching of mathematics, we need to do everything possible to ensure that all teachers are outstanding, which is why I believe the correct answer is E. Note, however, that recruiting and retaining outstanding teachers is only part of answer E. The other part is getting students to study harder. These two ideas go together, because great teachers are the ones who are best able to get their students to put in great effort.

Great Teachers Inspire Great Effort

In my "hint" a few paragraphs ago, I stated that my answer is the only one of the five options that we haven't really tried. This is clearly true: Research shows that in the world's best school systems, nearly all of the teachers are drawn from the upper echelons of college graduates. We have many similarly gifted teachers in the United States, but we also have substantial numbers who did not do so well in their own studies. Other research shows that American students by and large spend considerably less time studying (counting both class time and homework) than their peers in higher-performing nations. By the time they graduate high school, kids in many European and Asian nations have had the equivalent of one to two additional years of study time compared to American kids.

This is useful information, but did we really need research to tell us about it? Every single one of us has been to school, and many of us have kids in school. Don't we all *know* that the great teachers are the ones who taught us the most, and that the only way we ever truly learn something is by spending the time needed to study it?

Microsoft founder and philanthropist Bill Gates recently wrote a column about school reform for the *Washington Post* in which he stated that "of all the variables under a school's control, the single most decisive factor in student achievement is excellent teaching. It is astonishing what great teachers can do for their students."[30] What I find even more astonishing is that some people have tried to argue this obvious point with him. Even worse, some of those people have real power in the educational establishment. That is why, for example, a great program like Teach for America still has trouble getting many school districts to "accept" its teachers. Great teaching is in everyone's best interests, including the best interests of the teachers' unions, which would have far more popular support if they showed the same willingness to promote excellence and professionalism as they do to promote job security.

On the study side, I frequently give talks about strategies for math and science teaching, and I always start out by reminding the audience that the only way to learn is by studying. A teacher's job is not to pour knowledge into students' heads, but rather to help students make the effort to study and learn for themselves; as William Butler Yeats eloquently put it, "Education is not the filling of a pail, but the lighting of a fire." Despite this seemingly obvious fact, the evidence suggests we're moving in the wrong direction on the importance of studying. There's been a parental backlash against homework, and surveys show that college students today spend far less time studying than their peers of the past. In my own field of astronomy, a large number of people have been working to find ways to improve college teaching, but while their effort has led to some wonderfully innovative ideas for making effective use of class time, it has almost completely neglected the importance of getting students to study outside of class. I've even seen "research" papers trying to argue that effective use of class time may eliminate the need for outside reading and studying. Clearly, some of these people have lost sight of the forest for the trees.

Great teachers inspire their students to make the great effort of studying that is required to learn and to discover new things. There's nothing magic

---

30. Bill Gates, "How Teacher Development Could Revolutionize Our Schools," *Washington Post*, Feb. 28, 2011.

about it, and as a species, we've been teaching successfully for thousands of years; that is why, for example, we have computers and the Internet and the ancient Greeks didn't. The only real change has been in who gets taught. In the past, only a small fraction of the population received a formal education, which meant that nearly all teaching was either one-on-one or in small groups. Today, we believe that everyone has a right to be educated, which means we are attempting to mass-produce the teaching experience that used to belong to only a few.

There are great challenges to mass production, and educational research has turned up many useful insights that have helped educators develop better teaching strategies. But I will assert that two basic facts will never change because they are too fundamental to the way our brains work: (1) Great teachers are needed for great teaching; and (2) You can only learn by studying. As a nation, we ignore these facts at our peril, because as the second quote at the start of this epilogue says, "facts do not cease to exist because they are ignored."

Elementary School Math

The above two facts about teaching and studying apply to all subject areas. But in keeping with the theme I introduced in the first chapter, I'll now offer a few more specific thoughts on how we can cure our national propensity to be "bad at math." I'll proceed in order of educational level, starting with elementary school.

In my opinion, the goals for math in elementary school should be very simple: We need to make sure that kids form the foundation they'll need for more advanced math later, and we should ensure that they see math much as they see reading—as a tool that is useful for life, not just for a single subject area. Any curriculum that anyone develops should be judged against these two goals. Beyond that, there are a few general points that must still be addressed even if we have ideal curricula. Here is my personal list:

Skills are critical. A student who has not yet built conceptual understanding when he or she leaves elementary school can still build it later, but a student who hasn't built basic skills (such as addition, multiplication, fractions) is unlikely ever to catch back up. We must ensure that all kids do as much "drill and kill" as necessary to learn their basic skills. For example, unless a child has a learning disability, there's just no excuse for allowing him or her to reach fourth grade without knowing the multiplication tables, or fifth grade without being able to add and subtract simple fractions. When we allow students to miss out on such skills, we are very likely dooming them to a life of being "bad at math."

Provide plenty of study time. If we're going to match the results of higher-performing nations, our kids need to have as much study time as kids in those places. There are two ways to get more study time: We can either have our kids spend more time in school, or have them do more work at home (or some combination of both). In principle, I think that either option would be fine, but here's a practical reality: Today, kids from well-off families with educated parents almost universally get substantial help with math outside of school. At a minimum, they get help from their parents. Those who are struggling often get supplemental courses from programs such as Kumon; some even get personal tutors. Those who are doing well often get supplemental math instruction at home; for example, I've had my own kids do a wonderful online program offered through Stanford University, and we have friends who have their kids working through the Singapore math curriculum or the "JUMP math" curriculum at home. Poorer families can't afford these kinds of programs, and the kids with less educated parents don't have the support structure to help them. For these practical reasons, I believe that as a nation we need to dramatically increase the number of days our kids spend in school each year and the number of hours they spend in school each day. Summer vacation is great for those of us who can afford to send our kids to camps, enroll them in special programs, or take them on great trips, but for too many other kids it's just a time to forget what they've learned and fall even farther behind their peers. Perhaps there's a way to please everyone, such as by offering longer days and summer school as options rather than requirements, but one way or another, we must make sure that *all* kids have the time they need to study.

Set high but realistic expectations on an individual basis. Even if all kids had access to the very best programs, some would learn math slower or faster than others. This is OK—someone who learns more slowly won't necessarily learn less in the end. But it means that we have to recognize and accommodate different rates of learning. Kids will rise to meet high expectations, as long as those expectations are realistic. Low expectations will bore them, while unrealistic expectations will frustrate them. We therefore must allow teachers the flexibility to work with individual students at the rates that make sense for those students. This approach is already implemented at the middle and high school levels, where kids of the same grade take different math classes depending on how far they have advanced. But for some reason, it meets resistance in elementary schools, to the detriment of both the slow and the fast learners. The slow learners get labeled "bad at math," and, well, you know where that leads. The fast learners find themselves bored, which often leads them to dislike math, which can even make some of them become "bad at math" by the time they are adults.

Don't leave math in solitary confinement. There needs to be some time devoted to math each day in elementary school, ideally at least an hour. But that should not be the end of it; we need to integrate mathematical ideas throughout the curriculum, in much the same way that we've done it in this book. Just as we expect students to read in all their subjects, we should also expect them to make use of math. There's no reason why art can't be taught along with a bit of geometry, or music taught along with a bit about the mathematics of scales, or social studies taught along with some work on reading and making maps. Math is everywhere in our lives, so it should be everywhere in our education. Yes, some time will be devoted specifically to math, just as some time is devoted specifically to reading. But don't leave either of them confined to those single class periods alone.

All elementary teachers should be good at math. As we've discussed, none of the above ideas will make much difference unless we have outstanding teachers to implement them. Part of being an outstanding teacher is knowing your stuff. We wouldn't accept an elementary school teacher who only reads at a ninth-grade level, yet there are substantial numbers of elementary school teachers who couldn't pass a ninth-grade algebra test and would have a hard time discussing most of the topics in this book. I've even heard some teachers proclaim themselves to be "bad at math."

Clearly, if we want our kids to be good at math, we can't have teachers who think it's OK to be bad at it. Let's make sure that we draw teachers from the tops of their classes and that they are strong across *all* subject areas—and let's give them the pay and respect that they deserve for filling such a vital function in our society.

Middle School Math

Everything that I've said about elementary school math also applies to middle school math. The goals are the same, and the general principles are all the same. The only real difference is the level of material. In middle school, some kids will be ready to learn algebra, while others may need additional work at pre-algebra skills. Either way, we should be sure they learn all their mathematical skills and concepts well enough that they are prepared to move on to the next level. I'll add just three more specific notes.

Don't forget the importance of a context-driven approach. As we discussed at the end of Chapter 9, much of current math teaching is done backward. Teaching should always move from the concrete to the abstract, not the other way around. If we can implement this approach at the middle school level, kids will find math much more meaningful and enjoyable— which means they'll be more likely to succeed.

Math is not just for math teachers. A major difference between elementary and middle school is that the latter begins the process of specialization. In elementary school, the classroom teacher generally teaches all subjects. In middle school, math classes are taught by a math teacher who rarely teaches anything else. Because we still want students to see how mathematics is involved in all aspects of our lives, we need teachers in other subjects to use math wherever they can, which again means that they need to be good at it themselves.

Practice makes perfect. I'll hammer once more on the critical fact that you must study to learn. Let's say that a teacher has just covered some math topic and asked a question to see if students understand it, and they do. It's tempting to say that the students are "done" with that topic, but we all know that it won't stick unless they practice it much, much more. I know it can be tedious to be assigned fifty problems that all cover the same basic idea, and both kids and parents are likely to complain about it—but study and practice are the only ways to succeed. Consider the analogy to sports or music. It can take dozens or even hundreds of practice sessions to learn a new sports skill or play a new music piece, and even then you need to keep practicing—essentially keeping yourself in shape—if you are to retain it. Math works the same way. You need to drill the same ideas over and over for them to take hold, and then keep working at them to make sure you don't forget them. The amount of practice needed can vary individually, but my personal guideline is this: In middle school, I believe kids should be given enough math homework to keep the good students busy for at least a couple hours outside of class each week (which can include study hall, for example, if we go to longer school days). The slower students will take somewhat longer, but this is still not an unreasonable amount of math homework. Moreover, if someone can't complete the work in a reasonable amount of time, we should take it as a signal that the student needs additional help.

High School Math

Once again, all the themes I've described for elementary and middle school carry over to high school. So with the presumption that you've already read (and hopefully accepted) the principles I've described for those levels, I'll add three more that are specific to high school.

Four years of high school math for everyone. The reason we have four years of high school is presumably that we think it takes that long for students to learn everything we want them to know as high school graduates. Given that math is at least as important to modern society as any other subject, the implication is that we should make sure that all students take math in all four of their high school years. After all, we wouldn't say, "This year you don't have to read anything," so we should not say the equivalent for math. Of course, different students may take different levels of math classes, but everyone should take some kind of math every year.

Help kids keep their career options open. High school students often like to think of themselves as being adults or nearly so, but they are still kids, and we are still responsible for them. So consider this simple fact, which I also noted in Chapter 1: On average, careers that require advanced mathematical work—such as most careers in science or engineering—pay better and offer better working conditions than careers that don't. (Or, as it was put by famed teacher Jaime Escalante, subject of the movie *Stand and Deliver,* "Math is the great equalizer.") If high school kids try to tell us that they're not interested in those types of careers, we should still encourage them to at least complete algebra, so that all career options will still be open to them when they get to college. After all, they might change their minds as they get older. In my own teaching of astronomy for nonscience majors, for example, I've had dozens of students who were sufficiently inspired to tell me that they now wanted to study physics or astrophysics, but their mathematical preparation was so weak that it would have meant two or more years' worth of remedial work. A handful did it anyway, but the rest saw a dream slip away because they had not been properly prepared by their high school math experience.

Prepare high school graduates for modern life. If we set high expectations and teach well, the vast majority of high school students will be able to finish algebra well before their senior year. Some will be ready to move on to pre-calculus or calculus classes. Others may be better served by courses offering additional practice at other mathematical skills. Either way, however, we should encourage high school students also to take courses that will expose them to quantitative thinking in a broader sense, so that they will be able to understand the types of "math for life" topics that we've discussed in this book. After all, by now I hope I've made a convincing case that many of the problems we face as a nation and as a civilization are traceable to our having neglected to teach these kinds of skills in high school and college. With four years of high school math, there should be plenty of time to devote a semester or a year to a course in statistical or quantitative reasoning. Let's work to make these types of courses a standard part of the curriculum.

College Math

The issues change a bit once students enter college, because they generally are adults and therefore are responsible for their own choices. By and large, this means that college students split into two groups: those who have made the decision to pursue a major or career that will require calculus or more advanced math, and those who have not.

The first group is often known to educators by the acronym STEM, which stands for science, technology, engineering, and mathematics majors; personally, I prefer to think of them as the students on the calculus track. This group is easy for colleges to work with, at least in principle, because college mathematics departments have been built around the calculus track for decades. The major problems that colleges face in working with these students are that we'd like to have more of them to begin with (since careers in fields requiring advanced mathematics will be in high demand) and that too many current students drop off this track. As at lower levels, I believe that a little rethinking of the way we teach calculus and other advanced math classes could go a long way, particularly if we move to a more context-driven approach and make better use of technologies that can help students visualize mathematical ideas.

The more difficult challenge for colleges comes in deciding what to do with the students who have already made the choice to leave the calculus track. Most colleges require all students to complete at least one college course in mathematics, which is a very good thing given the importance of mathematics to our lives. Unfortunately, I believe that most colleges are still wasting this opportunity by teaching students something other than the "math for life" that they'll really need. The important point is that, for the vast majority of these students, their single required college math course will be *the last math course they ever take in their lives*. We therefore owe it to these students—and to the nation and world—to make the best possible use of the time the students will spend in this last math course. With that in mind, I'll offer a few specific suggestions.

Make college algebra an oxymoron. Nationwide, the majority of students who are not on the calculus track currently fulfill their college mathematics requirement by taking a course in "college algebra." This is pointless, for at least two reasons. First, most college students already have taken at least two years' worth of algebra in middle or high school; if it hasn't already sunk in, it's difficult to believe that one last semester of it will make a huge difference.[31] Second, students who don't plan to take more advanced math will never again use most of what we teach in algebra. Let's recognize "college algebra" for what it really is: high school algebra that is taught in college. As such, it should be considered a remedial course for those who need it because they hope to move on to more advanced math courses. As a remedial course, it should not count toward any graduation requirement.

Focus on quantitative reasoning. If you accept my rationale for no longer allowing algebra to fulfill the college math requirement, then the question becomes what to replace it with. To me, the answer is clear: quantitative reasoning. As you've seen throughout this book, most of the mathematical skills needed for quantitative reasoning are fairly basic, but the level of conceptual thinking can be quite advanced. This means that quantitative reasoning can be taught at a clearly collegiate level, and there is plenty to cover in a semester- or even a yearlong course; this entire book contains only about 5% to 10% as much material as a quantitative reasoning course typically covers. Moreover, because quantitative reasoning is so important to modern life, I believe it is a great disservice to make the requirement anything else. For example, some colleges have recently introduced course requirements in financial literacy, while others offer courses in statistical literacy; both types of course are clearly useful, but neither covers the breadth of topics that we've covered in this book, which means they are not by themselves enough. (Note, however, that such courses can be great options for one semester of a two-semester quantitative reasoning requirement.) Still other colleges offer courses giving students a brief introduction to some of the esoteric branches of mathematics that mathematicians study. These courses can be immensely interesting, but I don't think they are covering the material that students need for their

---

31. I once heard an algebra textbook author answer a question about the difference between "high school algebra" and "college algebra" approximately as follows: "The difference is simple. In college algebra, we teach students the same things that we taught them in high school algebra, only this time we teach it to them LOUDER."

everyday lives; for that reason, I'd make such courses electives, to come after a quantitative reasoning requirement is fulfilled.

You can't learn if you don't study. The single biggest problem in college mathematics education is the same problem that is harming all college education: the downward spiral in how much students are studying. Surveys show that the average number of hours that college students study outside class has fallen from about 25 hours each week in the 1960s to about 14 hours today. Unless you believe that students of today study much more efficiently than students of the past—and given the distractions that students now face from their electronic devices, it's far more likely that the opposite is true— then this dramatic reduction in study time can only mean that college students today are learning much less than their counterparts of the past. While it's easy to see the pressures on college faculty that have led to these reduced expectations of students, it's equally easy to see how detrimental this fact is both to students and to society at large. The solution, of course, is for colleges to institute policies to ensure that all courses require students to put in a traditional level of collegiate effort, which means two to three hours of study outside class for each hour in class. This solution admittedly will be difficult to implement in practice, but if we don't implement it, then college will increasingly become a waste of time and money for everyone involved.

Math for Life

What happens after college? I hope that I've convinced you that mathematics—especially the parts that qualify as quantitative reasoning—is just too important to be ignored during the rest of your life. As I've argued, problems from the housing bubble to the insanity of the federal debt and of current energy policies all stem from poor mathematical thinking. The only question is how to change this.

Toward that end, I'll offer a modest proposal for a three-step solution:

1. We should all regard being "bad at math" as a disease that must be cured. Don't let anyone get away with saying it with pride, or feeling that it is in any way acceptable.

2. Institute reforms in school and college curricula, such as the ones I've described in this epilogue, to help ensure that future high school and college graduates are better equipped to deal with the mathematics they will encounter in the modern world.

3. *Think* through the mathematics of every issue that we face. As we've seen in this book, the general outlines of solutions are often fairly obvious once you understand the real nature of the problems.

Please look back at the H. G. Wells quote with which I've opened this epilogue. It expresses the sentiment that drives me in my own work, and that I hope will drive you as well. We are indeed in a race between education and catastrophe, and I fear that we are falling behind. But I do not believe it is too late to make a comeback, and win the race upon which the futures of our children and grandchildren depend.

To Learn More

I hope that you will be inspired to want to learn much more about the role of mathematics in our lives than I've been able to cover in this short book. There are many resources out there to help you, from outstanding books and articles by other authors to the vast resources that you can find on the Web. Here I'll list just a few resources of my own that may be of interest to some readers:

- If you are interested in a full course on quantitative reasoning, either for self-study or because you are a teacher seeking to implement one, I hope you will consider my textbook written with Bill Briggs: *Using and Understanding Mathematics: A Quantitative Reasoning Approach*; the book is currently in its fifth edition, published by Addison-Wesley (a division of Pearson Education).
- If you want a more in-depth treatment of statistical reasoning, Bill Briggs and I have also collaborated on a textbook with Mario Triola called *Statistical Reasoning for Everyday Life*, also published by Addison-Wesley.
- Although I won't promise too much, I will try my best to post additional information and resources on the Web site for this book:

www.math-for-life.com

Acknowledgments

Although I am listed as the sole author of this book, its content is the result of a great collaborative effort in which many have participated. My own interest and involvement in quantitative reasoning dates back to the 1970s, when I was fortunate to be involved with developing math and science curricula at the elementary and middle school levels, while also working for a fantastic math tutoring program at the University of California, San Diego. During the 1980s, I was invited to participate with a faculty committee in helping create the University of Colorado's core requirement in quantitative reasoning, which to my knowledge represented the first formal effort to define the term "quantitative reasoning"; this committee then placed enough faith in me to let me develop a course curriculum for quantitative reasoning based on the new definition. I had plenty of help in that effort from too many people to name, but I wish to especially acknowledge J. Michael Shull, who served as a dean at the time and provided critical support to our curriculum development efforts, and several of my super teaching assistants, including Megan Donahue (now a coauthor of my astronomy textbook), Hal Huntsman, John Supra, Dave Theobold, David Wilson, and Mark Anderson. I benefited from many discussions with Cherilynn Morrow, who took over teaching of the quantitative reasoning course after I left. Most important, over the past twenty years I have worked closely with Bill Briggs to continue development of a quantitative reasoning curriculum, primarily by working jointly in writing a quantitative reasoning textbook.

All the above effort might have been for naught if not for the great efforts of my textbook publisher, Addison-Wesley. It signed the quantitative reasoning project at a time before the course existed outside the University of Colorado, then provided the editorial guidance necessary to help the book take a shape that would allow other colleges to institute similar courses. Many editors there have played critical roles, but I'd like to especially acknowledge Bill Poole, Greg Tobin, Anne Kelly, and Marnie Greenhut. Thanks to their efforts, our quantitative reasoning curriculum is now in use at more than two hundred colleges and universities. I also thank Addison-Wesley for allowing me to adapt numerous examples and illustrations from my textbook for inclusion in this book.

For helping me bring this book to fruition, I especially thank my dear friend Joan Marsh and my wife, Lisa, both of whom reviewed all of the chapters in draft form to help make sure the narrative was on track. Several reviewers also provided valuable feedback, including Shane Goodwin, Dave Taylor, Eric Gaze, and Rob Root. And I thank the publisher of this book, Ben Roberts, for believing both in this project and in me.

Finally, and as always, I thank my wife, Lisa, and my children, Grant and Brooke, for their patience and support in dealing with a sometimes ornery author.

Also by
Jeffrey Bennett

For Children

Max Goes to the Moon

Max Goes to Mars

Max Goes to Jupiter

The Wizard Who Saved the World

For Grownups

On the Cosmic Horizon: Ten Great Mysteries for Third Millennium Astronomy

*Beyond UFOs: The Search for Extraterrestrial Life
and Its Astonishing Implications for Our Future*

High School/College Textbooks

Using and Understanding Mathematics: A Quantitative Reasoning Approach

Statistical Reasoning for Everyday Life

Life in the Universe

The Cosmic Perspective

The Essential Cosmic Perspective

The Cosmic Perspective Fundamentals

Index

Index of Examples